THI E

JAPAN  **45**

TERRY C. TREADWELL

D1350331

Dedication:

For Toby, Sam and Rex.

Thanks to my wife Wendy for editing and proofreading the manuscript.

First published 2010

Amberley Publishing
Cirencester Road, Chalford,
Stroud, Gloucestershire GL6 8PE

www.amberleybooks.com

British Library Cataloguing in Publication Data.
A catalogue record for this book is available from the British Library.

ISBN 978-1-4456-0226-4

Typeset in 10pt on 12pt Celeste Pro.
Typesetting and Origination by Amberley Publishing.
Printed in the UK.

Contents

Introduction

This book does not claim to cover all of Japan's military aircraft from its nineteenth-century conception up to the end of the Second World War. The aircraft covered are those that were designed and put into production by the Japanese during this period. In the early years, almost all Japanese aircraft were either imported from Europe or America or were built in Japan under licence. To attempt to cover the complete range would run into several volumes.

The history of the companies involved in the design and manufacture of the Army and Navy aircraft is also covered. The intense rivalry between the Army and the Navy becomes very apparent, almost culminating in serious problems within Japan itself before it is brought under some semblance of control. But even during the Second World War the underlying animosity between the two services became a cause for concern, with each blaming the other for failures. Despite this the Japanese aviation companies did produce some superb aircraft, none more famous than the Mitsubishi A6M Zeke (Zero) and the Aichi M6A Sierans that were launched from the I-400 class submarines. Then there was the Yokosuka E14Y1 (Glen), built by the Yokosuka Arsenal and launched from the I-25 submarine. This aircraft, flown by Warrant Officer Nobuo Fujita, carried out the only attacks on the American mainland when it dropped incendiary bombs on the forests of Oregon.

In the early years of aviation in Japan, the Army and the Navy had their own aircraft identification system. The Army, because they used almost exclusively western aircraft like the Nieuport, Sopwith and Maurice Farman, devised a way in which the Japanese could identify them. For example, using the English alphabet, the letter A was given to the Nieuport and was pronounced 'Ko' by the Japanese, so a Type A 4 Nieuport became Type Ko 4.

The Navy, however, confused everybody by using three different designation systems to identify their aircraft over a six-and-a-half-year period. These took the form of the Japanese katakana characters of I, Ro, Ha, Ni, followed by the suffix 'go' (type). In some cases, in an effort to further identify an aircraft that had been modified, the designations were followed by 'gara' or 'kata', which meant sub-type or model, i.e. Ko-gata (A Model), Otsu-gara (B Model).

In 1927, the Japanese system of numbering aircraft was changed and based on the year AD 1925, which is the Japanese year 2585. Take, for example, the Type 98 aircraft: the last two digits of the Japanese year, 2598 (1938), in which the aircraft was accepted, would be the designation given to it. When the Japanese year 2600 was reached, the Zeke fighter became the Navy Type 0 Carrier-Based Fighter. The Army, however, at this time used the designation Type 100, but the following year, 2601, it dropped back to using the single digit – Type 1. The term 'Shi' was also used with experimental aircraft and is a shortened version of the term *Shisauku Seizo* (trial manufacture), i.e. Experimental 7-Shi Reconnaissance Seaplane. This went some way towards simplifying the identification system.

Because Japanese aviation was relatively unknown, when the Second World War began there was no formula for identifying the various aircraft unless you had access to the complicated Japanese system, so the Allies decided to give them names. Up to this point all Japanese fighters were invariably called Zekes, and bombers were given the name of what was thought to be the manufacturer, i.e. Nakajima, Mitsubishi or Kawanishi. The giving of names to the various aircraft proved to be the one way of identifying them with relative ease. The way the names were selected was simplicity itself: they were just named after people. For example the sturdy Mitsubishi G4M1 was given the name 'Betty', it is said, after a rather well-endowed American nurse. An intelligence officer by the name of Captain Frank McCoy was one of those said to have come up with the idea, after unsuccessfully trying to identify various Japanese bombers.

McCoy came from the state of Tennessee, and he used hillbilly nicknames such as Zeke, Rufe, Jake and Nate as a means of identifying the aircraft. As more and more Japanese aircraft were noted, they were given names until an almost complete picture of the Japanese Air Arms, both Army and Navy, had been identified by Western names. Some of the new aircraft types were named after the pilots or crewmembers that first spotted them and could give detailed descriptions of what they saw. Flight Sergeant George Remmington RAAF, a member of the Allied intelligence team, used all this information and rapidly turned the descriptions of these aircraft into silhouettes. In fact, the Kawanishi N1K1-J Navy Fighter was given the code name 'George' after him. These silhouettes became the official means of identifying Japanese aircraft during the Second World War. By December 1942, almost all the existing Japanese military aircraft had been identified, given Allied code names and been accepted as such by all the Allied powers.

A lot of the material pertaining to the development of Japanese aviation has had to be acquired from a variety of literary sources, mainly because of the Japanese military's slack attitude towards maintaining records. After the Pacific War, almost all the records kept by the Japanese military were either destroyed or written in such a way that they bore no resemblance to the truth. The latter was a deliberate attempt to shield the Japanese people from what was really happening towards the end of the war.

CHAPTER ONE
The Early Years – 1877-1919

It was the uprising in the Kyushu area of Japan in 1877 that heralded the birth of Japanese aviation. The rebellion concerned the Satsuma clan samurai, who revolted against the Japanese Imperial Army. The government forces were besieged in the Kumamoto Castle, unable to get messages out for reinforcements to be sent. Then it was remembered that during the siege of Paris in the Franco-Prussian War of 1870-71, the beleaguered French used man-carrying balloons to convey messages to the outside world. The Japanese government ordered the making of two of these types of balloons, and two naval engineers, Shinpachi Baba and Buheita Azabu, set to work to construct them. The first balloon was constructed by Shinpachi Baba, filled with coal gas and achieved a height of 715 feet (218 m), becoming the first manned balloon flight in Japan. Fortunately reinforcements arrived before the balloons could be used and the siege situation was resolved. The use of the balloon was further highlighted when Second Lieutenant Shinroku Ishimoto attained a height of 300 feet in a balloon he had built himself.

Initially the government's military departments were interested in the balloons from a reconnaissance point of view, but when the uprising was subdued by conventional methods, they shelved the idea. From the very onset, it is clear that the Japanese military were leading the way as far as aviation was concerned and would continue to maintain control over the design and manufacture. The Japanese Army, however, still considered the balloon to be of importance and ordered a gas-filled balloon, together with a mooring car and gas generator, from the French balloon maker Gabriel Yon.

Then, in 1904, when the war with Russia erupted and the use of heavy artillery became part of modern warfare, it was decided to use the balloon for both reconnaissance and artillery spotting.

An order was placed with Isaburo Yamada, a known balloon maker, for two balloons capable of carrying a man. An additional balloon was ordered from Charles Spencer, an English balloon-maker who was travelling in Japan at the time. Spencer made a number of flights over Tokyo in his balloon and also made a number of parachute jumps, which thrilled all the spectators and intrigued a number of military observers.

The balloons bought from Spencer and Yamada were used for both reconnaissance and artillery spotting at the Battle of Lushun, also known as the invasion of Port Arthur. The success of the battle gave impetus to the development of the balloon as a tool for the military.

This was taken one step further when, in 1909, an Englishman by the name of Benjamin Hamilton arrived in Japan to demonstrate his airship. His first airship flight at Ueno Park, Tokyo, drew large crowds, and among them was Isaburo Yamada. He had been working on the design of an airship and, after seeing how successful Hamilton's airship was, decided to use the design as the basis for his.

In September 1910, the first test flight of Yamada's airship ended up in the grounds of a military base because of a hydrogen gas leak. An explosion destroyed the second in 1911, but the third, in the same year, carried out a number of flights around Tokyo. The military, although impressed, were not impressed enough to place an order and the airship was eventually sold to China.

THE DEVELOPMENT OF HEAVIER-THAN-AIR CRAFT

In America and Europe a new development of the lighter-than-air craft was gathering momentum. The Wright Brothers had carried out the first powered flight of a heavier-than-air craft, which was closely followed by more powered flights taking place in Europe. Back in Japan, the military established a new section to their military arm, the PMBRA (Provisional Committee for Military Balloon Research [Rinji Kikyu Kenkyu Kai]), with Lieutenant-General Gaishi Nagoka at its head. The development of aircraft in America and Europe was quickly gathering pace and Japan was quick to recognise this. Although the title of their section was balloon research, General Nagoka's committee concentrated their efforts on the development of powered aircraft and sent two Army captains, Yoshitoshi Tokugawa and Kumazo Hino, to France and Germany respectively to be taught to fly. On completion of their training, the two new pilots purchased a Henry Farman biplane and a Hans Grade monoplane respectively. On their return to Japan both pilots demonstrated their aircraft in front of huge crowds at Yoyogi Field, just outside Tokyo. The military hierarchy was impressed and immediately saw the benefits of the aeroplane in a military role. In 1911, the Army established a military airfield at Tokorozawa, just outside Tokyo, and within a year five more Army pilots had been trained.

Initially this had been a joint Army/Navy venture, with the Army being the prominent member. But such was the rivalry between the two services that the Navy decided to go it alone, mainly because they wanted overall control of their own aviation section, and up to this point only Army pilots were being trained. In June 1912 the Japanese Imperial Navy split from the Army and created its own Naval Committee for Aeronautical Research (Kaigun Kokujutsu Kenkyu Kai). They immediately sent three lieutenants to France and three to America. One of the lieutenants, Yozo Kaneko, who went to France, returned with two Maurice Farman seaplanes. Lieutenant Sankichi Kohno, who went to America, returned with two Curtiss seaplanes. On their return, flights were carried out at Yokosuka on 2 November 1912, the first powered flights to be made by the Japanese Naval Air Service. The Curtiss aircraft was initially given the designation Navy Type Ka Seaplane; this was later changed to I-go Otsu-gata Seaplane.

THE FIRST AIRCRAFT COMPANIES ARE ESTABLISHED

First Lieutenant Chikuhei Nakajima, who had retired from the Navy to set up his own small aircraft factory, produced a small experimental seaplane in 1913, followed in 1915 by three Nakajima tractor biplanes, all for the Navy. The Kawanishi Shipbuilding Company had a subsidiary company called Nihon Aircraft, and they had financed Nakajima. The initial success of the Nihon Company drew the attention of the military and in 1915 they took over its running. The following year the Nakajima Aircraft Company appeared, while Kawashini established its own aircraft division within its own company. The Mitsubishi and Kawasaki companies appeared a year later.

The first recognised Japanese-designed aircraft were of the Kai-Shiki series. They were built in 1914 under the control of Army captain Yoshitoshi Tokugawa, one of Japan's first pilots. Two aircraft were built and both were flown by Captain Tokugawa but neither went into production.

The intense rivalry between the Navy and the Army was becoming a source of increasing concern for the Japanese government. The development of aviation and the possibility of setting up a separate air arm under the control of the Army was becoming a bitter pill for the Navy to swallow. The Army and the Navy continued to go their own way, and by 1914 they had a total of twenty-eight aircraft, sixteen to the Army and twelve to the Navy. All of these aircraft were either from Europe or America as no effort had been made to develop Japan's own manufacturing industry.

The first of the foreign aircraft to be produced for the Army were the Maurice Farman biplanes, built at a small factory at an airfield near the town of Tokorozawa in 1913. The

Navy commenced production of the Maurice Farman seaplanes at its torpedo plant at the Yokosuka Navy Arsenal at the same time. Both plants continued to manufacture Maurice Farman series aircraft until 1921. Other types of aircraft were being built under licence by two different arsenals in Tokyo; they included Sopwith 1½ Strutters, Nieuport 18s and SPAD S.11s, in addition to over fifty Sopwith Pups being imported from Britain for both the Army and the Navy.

THE FIRST WORLD WAR

At the onset of the First World War, the Japanese joined in on the side of the Allies and in the summer of 1914, using their air power, they attacked the fortress of the German-Austrian troops at Tsingtao. During the two-month operation the Japanese Army Air Arm carried out eighty-six sorties and dropped forty-four bombs, while the Naval Air Arm carried out forty-nine sorties and dropped 199 bombs. The prime target, the German cruiser SMS *Kaiserin Elisabeth*, was not touched but one torpedo boat was sunk. They also carried out a number of reconnaissance flights over the harbour and inlets in an attempt to locate the heavy cruiser SMS *Emden*, which they had been told was at anchor. The flight also confirmed the exact number of German ships that were at anchor, in addition to the troop movements in the area.

The missions showed the potential of the use of aircraft in a war situation and the value of reconnaissance flights over enemy-held territory. Both the Army and the Navy used the opportunity to build their own airpowers.

At the end of December 1915, the Army had created their first Air Battalion, which consisted of three squadrons, each containing four aircraft. By the middle of 1916 a second Air Battalion had been formed, and both battalions were becoming an integral part of the Army's regular manoeuvres. During the following two years the Army's Air Battalions trained with great intensity. Then, in 1918 Japanese troops were sent to Siberia after the murder of the Japanese consul general in Nikolaevsk-na-Amure, and both Air Battalions also went to support the ground troops. It was there that they gained invaluable experience in maintaining and operating aircraft under the most arduous and severe weather conditions.

The need for training aircraft became a priority and the Army looked towards a small private company known as Nihon Hikoki Seisakusho, who built small private aircraft. It had designed and built a two-seat trainer and had produced four prototypes, all of which had been tested by the Army and rejected.

The Army then turned to Chikuhei Nakajima's company, who were in the process of building aircraft. The first three prototypes were never put into production, but the fourth prototype was the one that the Army accepted. The Nakajima Type 5 Biplane was a two-seat, two-bay biplane of fabric-covered, wooden construction powered by a 165-hp Hall-Scott A-5a 6-cylinder engine. A total of 118 were built between 1919 and 1921. Of these, 100 went to the military; the remainder were sold to various civil organisations. The success of this aircraft prompted the military to look at it as a trainer, and after a number of modifications were made, a contract for twenty of the aircraft was awarded. In 1919 the first mass-produced aircraft came off the production line, the Nakajima Type 5 Trainer. This was a two-seat biplane of fabric covered, wooden construction, powered by a Hall-Scott A-5a 6-cylinder, water-cooled, in-line 150-hp engine.

The order was increased the following year to 100, an unusually large military order for an aircraft from a civilian company. The vast majority of aircraft at this time were being imported from the Western world. As the aircraft rolled off the production line they were being assigned to various Army flying units, but within months of the first of them being delivered a series of accidents occurred, resulting in some serious injuries and one or two deaths. It was discovered that stalls were being prematurely induced because the wing ribs were being made to incorrect drawings. With this problem quickly resolved, another one raised its head. A number of the aircraft suffered from engine fires after a long flight, and it was discovered that a build-up of engine oil in the bottom of the engine cowl had a tendency to catch fire. After a couple of incidents, including one when the pilot and his mechanic died, the Army decided to replace the trainer. By the end of 1921, 101 of these aircraft had been built for the Army.

In 1919, it was recognised that the training of pilots could be greatly assisted by learning from the French method. They had developed a ground-taxiing trainer based on a short-winged Morane-Saulnier fighter. The Hombu Kombu ordered the Tokorozawa Army Supply Department to build a similar trainer using the Nieuport 81 as the design, but with much shorter wings and a smaller engine. Given the designation Type Ni (Nieuport), ten of the trainers were built, and when accepted by the Army they were re-designated Army Model 2 Ground-Taxiing Trainer. The following year the Nakajima factory produced five more of the trainers.

During the First World War, the Yokosho Company manufactured a number of seaplanes. Amongst them was one designed by Lieutenant Nakajima and Lieutenant Umakoshi, which became known as the Navy Yokosho Ro-go Ko-gata Reconnaissance Seaplane. The first prototype took to the air at the beginning of 1918 and was immediately accepted by the Navy. Production started almost immediately and the first four rolled off the production line at the end of 1918. A 140-hp Salmson engine powered the first of the aircraft, but the 200-hp Mitsubishi Hi engine quickly replaced this.

At the beginning of 1919, three of the first four production models were converted to single-seat aircraft and used for long distance flights. The Yokosho Naval Arsenal carried out production of the first thirty-two aircraft, and then production was switched to the Aichi and Nakajima Companies, who between them built a further 186 between 1918 and 1924.

SPECIFICATIONS

Army Type Mo (Maurice Farman Type) 1913 Aeroplane

Wing Span:	50 ft 11¾ in. (15.5 m)
Length:	37 ft (11.3 m)
Height:	11 ft 3¾ in. (3.45 m)
Weight Empty:	1,280 lb (580 kg)
Weight Loaded:	1,885 lb (855 kg)
Max. Speed:	59 mph (51 knts)
Ceiling:	9,483 ft (3,000 m)
Endurance:	4 hours
Range:	Not known
Engine:	One 70-80-hp Renault 8-cylinder, air-cooled radial
Armament:	None

Nakajima Type 5 Trainer

Upper Wing Span:	41 ft 4¼ in. (12.6 m)
Lower Wing Span:	41 ft 4¼ in. (12.6 m)
Length:	23 ft 1½ in. (7.04 m)
Height:	9 ft 5½ in. (2.88 m)
Weight Empty:	1,719 lb (780 kg)
Weight Loaded:	2,491 lb (1,130 kg)
Max. Speed:	80 mph (70 kts)
Ceiling:	11,155 ft (3,400 m)
Endurance:	4 hours

Engine: 165-hp Hall-Scott A-5a 6-cylinder, air-cooled rotary
Armament: None

Nakajima Type 5 Biplane

Upper Wing Span: 41 ft 4¼ in. (12.6 m)
Lower Wing Span: 41 ft 4¼ in. (12.6 m)
Length: 23 ft 1½ in. (7.04 m)
Height: 9 ft 5½ in. (2.88 m)
Weight Empty: 1,719 lb (780 kg)
Weight Loaded: 2,491 lb (1,130 kg)
Max. Speed: 80 mph (70 kts)
Ceiling: 11,155 ft (3,400 m)
Endurance: 4 hours
Engine: 150-165-hp Hall-Scott A-5a 6-cylinder, air-cooled rotary
Armament: None

Navy Type Ka Seaplane (I-go Otsu-gata Seaplane)

Wing Span: 37 ft 2¾ in. (11.3 m)
Length: 27 ft 9 in. (8.4 m)
Height: 8 ft 2 in. (2.49 m)
Weight Empty: 1,180 lb (535 kg)
Weight Loaded: 1,642 lb (745 kg)
Max. Speed: 50 mph (43 knts)
Ceiling: 9,483 ft (3,000 m)
Endurance: 3 hours
Range: Not known
Engine: One 75-hp Curtiss O 8-cylinder, water-cooled
Armament: None

Navy Yokosho Ro-go Ko-gata Reconnaissance Seaplane

Wing Span: 50 ft 11½ in. (15.5 m)
Length: 33 ft 4½ in. (10.1 m)
Height: 12 ft 1 in. (3.6 m)
Weight Empty: 2,669 lb (1,211 kg)
Weight Loaded: 3,694 lb. (1,676 kg)
Max. Speed: 86 mph (75 kts)
Ceiling: Not known
Endurance: 5 hours
Engine: 200-hp Salmson 2M-7, 9-cylinder water-cooled, radial
Armament: One dorsal 7.7 mm machine gun

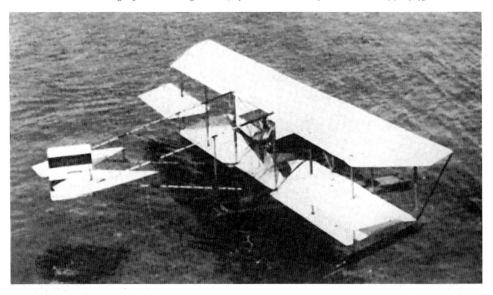

Navy Type Seaplane.

Japanese-built Henry Farman.

Nakajima Type 5 trainer.

Bemused onlookers looking at No. 3 Kai-Shiki.

No. 3 in the Kai-Shiki Series.

SMS *Kaiserin Elisabeth.*

Nakajima Type 5 biplane.

Army Model 2 ground-taxiing trainer.

Yokosho Ro-go Ko-gata seaplane.

Yokosho Ro-go Ko-gata reconnaissance seaplane.

Army Model 3 ground-taxiing trainer – hence the stubby wings.

Mitsubishi Army Type Ko 1 trainer.

Nakajima Army Type Ko 3 fighter.

Nakajima Ko 4 fighter trainer.

Nakajima Army Type 91 fighter.

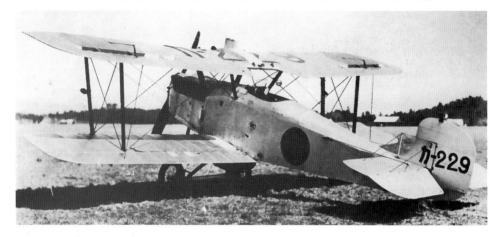

Mitsubishi Navy Type 10 carrier fighter.

Mitsubishi Navy Type 10 carrier fighter.

Maurice Farman Type Model 6 (Mo-6).

Kawasaki Army Type Otsu 1 reconnaissance aircraft.

Yokosho Yi-ko.

Mitsubishi 1 MT torpedo-bomber.

Nakajima Navy Type 15 reconnaissance seaplane on slipway.

Hiro Navy Type 15 flying boat (H1H1).

Kawanishi Navy Type 90-3 reconnaissance seaplane.

Yokosho 1-go reconnaissance seaplane, initially designed to be carried on a submarine.

Nakajima Navy Type 90-2-2 reconnaissance seaplane (E4N2).

Watanabe E9W Type 96 Model 11.

Kawasaki Army Type 92 (KDA-5).

Yokosuka B3Y1 Navy Type 2 carrier attack aircraft.

Kawasaki Type 88 light bomber.

CHAPTER TWO

1920S

Following the success of the Model 2 trainers, in 1921 the Tokorozawa Supply Department designed and built its own trainer, which was slightly smaller and lighter than the earlier models. Constructed of wood and covered in fabric, the Army Model 3 Ground Taxiing Trainer, as it was designated, was a shoulder-wing, single-seat monoplane powered by a 30-hp Anzani 3-cylinder, air-cooled radial engine. It was used to train student pilots to taxi an aircraft on the ground, as its name implies. A total of thirty of the trainers were built over the next two years.

FRENCH IMPORTS

The end of the war in Europe brought an influx of French aircraft into Japan. The Japanese Army bought a number of aircraft including the SPAD VII and XIII, Morane-Saulnier, Nieuport 24 C.1, 27 C.1s, 81 and 83-E2s. They also purchased fifty Sopwith Pups with the intention of bolstering their fleets of training aircraft and obtaining a licence to build copies in Japan.

The Nieuport 81 and 83-E2s became the standard Army trainers and were initially built under licence by the Army at Tokorozawa. It soon became obvious that the Army was not equipped to build aircraft, and the production of the 81-E2 was passed to the Mitsubishi company. Production of the 83-E2 was given to the newly formed Nakajima company. The design and construction of both these aircraft was identical to that of the French-built models. When completed they were given the Japanese designation of Ko 1 for the 81-E2 and Ko 2 for the 83-E2. A total of fifty-seven of these trainers were built between 1922 and 1924.

Among the other imported French aircraft were the single-seat biplanes, the Nieuport 24 C.1 and 27 C.1. A licence was obtained to build these aircraft and the 24 C.1 was built as a single-seat trainer, while the 27 C.1 was built as a fighter. This aircraft was found to be the most manoeuvrable of all the imported aircraft and was adopted by the Army as their standard fighter. A total of 102 of the aircraft were built by the Nakajima company under the designation Nakajima Army Type Ko 3 Fighter/Trainer.

In addition to gaining the licence to build the aircraft, the Army acquired the licence to build the Le Rhône engine. Initially the aircraft were given the designation Type Ni-24, the Ni being the first two letters of the Nieuport company, but in November 1921 a new system came in for the Army; the Ni became Ko and the aircraft became Ko 3. The Type Ko 3 Trainer was identical to the Nieuport 24 C.1 and the production models that were assigned to fighter units in 1921 became the mainstay of the JAAF until they were replaced in 1926 by the Nakajima Type Ko 4.

The success of the Army Type Ko 3 prompted the Army to develop one of the Nieuport 83 E2s as a trainer. The production of this aircraft started in March 1922 and ended in July the same year, by which time forty of the Type Ko 2 Trainers were built.

The Type Ko 3 was replaced in 1926 by the latest Nieuport fighter, the Nieuport 29-C-1. The fighter had seen action on the Western Front during the First World War, and was said to have been the best fighter aircraft in the world at the time. The Japanese Army purchased a number of these aircraft in 1923 and Nakajima acquired the licence to build the model in Japan.

The Army Type Ko 4 Fighter, as it was designated, was very different from the previous Nieuport models, inasmuch as the fuselage was of a streamlined, smooth-skinned, wooden monocoque construction and had an equal wingspan covered in fabric. When the first one rolled off the production line it was almost identical to the Nieuport 29-C-1, the only minor differences being the Lamblin radiator and the dihedral on the upper wing. Its armament consisted of two 7.7 mm Vickers machine guns mounted on top of the forward section of the fuselage. The first models came off the production line in 1925, were assigned to front line units and saw action during the incidents in Manchuria and Shanghai. The Nakajima Army Type Ko 4 was the Japanese Army's first mass-produced fighter aircraft and it remained in service from 1925 to 1933, when it was replaced by the Nakajima Army Type 91 Fighter. A total of 608 Army Type Ko 4s were built between December 1923 and January 1932.

ARMY REORGANISATION

The Army Air Corps officially came into being on 1 May 1925 and was placed under the command of Lieutenant-General Kinichi Yasumitsu. It had a strength at this stage of 3,700 officers and men, and a total of 500 aircraft at its disposal. Originally known as the Army Air Division, it had already experienced action in 1920 at Vladivostok and in 1928 was to see more action during the Tsinan Incident in China.

As the Air Corps grew, the need for reorganisation became more and more apparent. As Army Air Divisions, it comprised units made up of fighter, bomber, reconnaissance and transport squadrons known as Chutais. The experiences gained during the China conflicts showed that smaller, more specialised units were required; these became known as Sentais (Groups).

Under the new system, a Sentai was comprised of three Chutais of between nine and twelve aircraft and had its own Santai Hombu (Headquarters Section).

Towards the end of 1927, the Army decided to replace some of their fighter aircraft, the Ko 4, and asked the leading aircraft manufacturers, Kawasaki, Mitsubishi and Nakajima to submit proposals. This was the first time a proposal for a Japanese-designed aircraft had been subjected to a competition. All the designers from the three companies looked toward Western-designed aircraft for their inspiration. Mitsubishi and Kawasaki studied the German-designed fighters, while Nakajima considered the French designs.

It was the Nakajima company that produced the first prototype, a parasol-winged fighter powered by an air-cooled radial engine. The other two quickly followed this entry, but during dive tests of the Mitsubishi entry the aircraft came apart in the air, killing the pilot. The other two fared little better during their tests and the Army rejected all three designs on the grounds that they were not strong enough to withstand the rigours required for a fighter aircraft. However they decided that of the three, the Nakajima model might be worth pursuing.

The company was asked to modify its design so as to meet the requirements and specification laid down by the Army. Five prototypes were built between 1929 and 1931 with extensive modifications. During tests it was found that the centre of gravity was too far back and further design modifications had to be made to rectify it. Further tests convinced the Army that it had found its replacement, and the aircraft was put into production with the designation Nakajima Army Type 91 Fighter.

This was a single-engined parasol monoplane, with an all-metal fuselage and metal-framed wings covered in fabric, as were the tail surfaces. The first models off the production line came at a time when the Manchurian conflict started and were almost immediately in action.

THE SHANGHAI INCIDENT

The outbreak of the Shanghai incident caused the Type 91-1 to be pushed into its fighter role once again and production of the aircraft was increased. The second model to come off the production line, the Type 91-2 was powered by a 450-580-hp Nakajima Kotobuki 2 9-cylinder, air-cooled, radial engine as opposed to the 450-520-hp Nakajima Jupiter VII 9-cylinder, air-cooled, radial engine of the Type 91-1.

The installing of the new engine caused the front section of the aircraft to be changed. The fitting of Townsend ring cowling, which changed the shape of the nose, also enlarged the engine area. A total of 450 of the aircraft were built, but only twenty-two of them were Type 91-2s.

The JNAF was in need of a replacement for its Mitsubishi Type 10 Carrier Fighter towards the end of 1926, and so asked the three top aircraft manufacturers, Aichi, Mitsubishi and Nakajima, to submit proposals. Because it was still reliant to some degree on the Western manufacturers, Nakajima approached the Gloucestershire Aircraft Company in England to construct a modified version of their Gamecock Gamber aircraft. Nakajima sent one of its designers, Takao Yoshida, to England to work with the Gloucestershire company. The Gamecock aircraft had been built to operate from airfields, but the Nakajima company required its version to operate from aircraft carriers, so a number of modifications had to be made to strengthen the aircraft.

The wingspan of the Gamecock Gamber was increased from 9.18 m to 9.70 m, which increased the wing area by 2 sq. m. This improved the aircraft's manoeuvrability and increased its performance. The aircraft, designated the Nakajima Navy Type 3 Carrier Fighter (A1N1), was powered by a 420-hp Nakajima Jupiter VI, 9-cylinder, air-cooled radial engine. After a series of tests, the Navy accepted the aircraft in April 1929.

In 1930 the aircraft was fitted with a 460-hp Nakajima Kotobuki 2, 9-cylinder, air-cooled radial engine and given the designation A1N2. Two years later, when the first of the Shanghai Incidents erupted, the A1N2 was the first of the Japanese Navy's aircraft to go into combat. Three of the aircraft, from the aircraft carrier *Kaga*, shot down a Boeing P-12 flown by an American pilot, Robert Short. Between 1929 and 1932 a total of 150 A1N1s and A1N2s were built.

The Shanghai Incident occurred just after a number of anti-Japanese riots in Shanghai, when a number of Japanese citizens were killed. The Japanese Consul-General in China demanded that the mayor of Shanghai, General Wu Te-chen, suppress all anti-Japanese propaganda and punish the people responsible for the deaths of the Japanese citizens.

Before this could happen, the commander of the Japanese Yangtze Squadron launched an attack on the city on 28 January 1932. Fierce fighting then ensued between Japanese and Chinese forces, including the first aerial attacks by Japanese aircraft. During an attack on Hangchow airfield by Nakajima A1N2s from a land base near Shanghai, Japanese naval aircraft were involved in an aerial battle with Chinese aircraft and scored a number of victories.

The brief war was brought to an end through the intervention of the British government and the League of Nations in May 1932, but the gradual expansion of Japanese forces in China continued unabated.

This incident prompted Japanese aircraft manufacturers to increase their design and production of military aircraft. One of these was the Kawasaki company, who had not been idle and had produced another fighter known as the Kawasaki Type 92, and this replaced the Type 91 the following year. The Nakajima Type 91 was still regarded by many pilots as being superior to the Type 92, but its role was reduced to that of long-range operational flying training.

NAVY REORGANISATION

In April 1916 the Japanese Navy had formed their first Air Corps (Kokutai) at Yokosuka, followed in 1917 by the Second Air Corps at Sasebo. It continued to build on its fleet of aircraft, and by 1920 it had created three squadrons out of the two Air Corps. The squadrons consisted of two combat squadrons, each with six aircraft, and one training squadron with eighteen aircraft. Both the Army and the Navy had purchased foreign aircraft because there were no aircraft manufacturers in Japan at the time. This is not to say that there were never Japanese-built aircraft. In 1909, Army Lieutenant Kumazo Hino and Naval engineer Baron Sanji Narahara had designed and built their own aircraft. When they flew in May 1909, they were the first powered flights by Japanese-designed and -built aircraft. Neither was ever considered seriously by the military.

Aircraft continued to be designed and built in Japan, but almost all of them were unsuccessful, and either did not get past the design stage or failed the flight tests. It was then decided that European and American aircraft were far superior in both design and construction, so a decision was made to build the aircraft in Japan under licence.

Although the Japanese were still adopting European designs for their aircraft, they were starting to adapt them to their own specifications. One of these was the Army Maurice Farman Type Model 6 (Mo-6) that had been built in 1916. It was almost a copy of the Maurice Farman Mo-4 that had been imported from France, the difference being that it was heavier and larger overall and had coolant radiators mounted on either side of the engine. After tests the aircraft was put into production; the production models, however, had shorter front skids, which meant that the diagonal strut in the front, which supported the longer version, was no longer required. The first few off the production line impressed the Army, but during a series of Army manoeuvres on the Ohmi Plain near Lake Biwa, where the aircraft was put through its paces under simulated battle conditions, twelve of the fourteen aircraft taking part either crashed or had to make emergency landings. An investigation committee of engineers and pilots was hurriedly convened in an effort to find out the cause of so many mishaps.

The result was that improved engines were called for, better materials were to be used in the building of the aircraft, and engineers had to have more training. Production was halted for a while the improvements were implemented, resulting in much-improved aircraft that had an extended life as an operational aircraft.

This was also to be the last of the Army's pusher biplanes and became the primary aircraft for the First Army Air Battalion, used for reconnaissance and training purposes until 1923. Between 1917 and 1921, 134 of the aircraft were built.

The Navy was also starting the production of aircraft when, in 1917, Lieutenant Kishichi Umakoshi and Lieutenant Nakajima designed a single-engined scout biplane seaplane. The Ro-go-Ko-gata, as it was called, was built by the Yokosho arsenal as a reconnaissance seaplane and was the most successful aircraft of the day. A total of 218 were built over the next seven years, making it the main aircraft of the Navy. It was a single-engined, twin-float, two-seat biplane of wooden construction covered with fabric. The 50 ft 11½ in. (15.53 m) wings folded back against the 33 ft 4½ in. (10.17 m) fuselage for ease of stowage. Initially a 140-hp Salmson M9 engine powered the aircraft, but a 200-hp Mitsubishi type Hi engine later replaced this. Three of the aircraft were later converted to single-seaters, giving them a greater range, as they were able to carry additional fuel.

Such was the demand for the aircraft that a number of companies were involved in its production. A total of 218 were built: thirty-two by the Yokosho company, eighty-two by the Aichi company and 106 by the Nakajima company.

Aircraft were not the only thing being built under licence: so were the various engines. The Japanese imported a wide range of engines: from France, the 70- and 100-hp Renault, 50 and 100-hp Gnome, 60-hp Anzani, 100-hp Clerget, 80 and 100-hp Le Rhône, Salmson 150 and 200-hp and 150 and 200-hp Hispano-Suiza; from England the 150-hp Sunbeam; from America the 75 and 90-hp Curtiss; and from Germany the 100-hp Daimler Benz. With the exception of the Le Rhône and the Clerget engines, the remainder were built under licence in Japan.

A NEW TRAINING REGIME

With the military now convinced of the value of aircraft, the need for training became a priority. With the exception of minor skirmishes, the Japanese pilots had no experience whatsoever with regard to deploying their aircraft as a weapon of war. It was decided to invite European pilots to Japan to instruct the pilots, and once again the rivalry between the Army and the Navy manifested itself. In 1919, the Army had invited sixty-one French officers and men from l'Armée de l'Air under the command of a Colonel Faure to instruct their aircrews on aerial combat, gunnery, reconnaissance and bombing. The Navy immediately invited the British to send instructors to Japan to instruct its crews. Under the command of Captain Sir William Sempill, RN, twenty-nine instructors arrived at the Navy's first purpose-built air base at Kasumigaura. Over the next year they taught the Japanese pilots how to carry out torpedo bombing, reconnaissance, gunnery spotting, aerial photography and a number of other operations required by the Navy.

The results from these programmes were to alter the structure of Japanese military aviation considerably. The Japanese Army changed its administration system by abolishing the PMBRA that up to this point controlled all aviation-related matters, and replaced it with an Aviation Section (Kokubu) within the Ministry of the Army. They created six Air Battalions, each one consisting of three squadrons.

The First and Second Air Battalions were based at Kagamigahara, the Third Air Battalion was based at Yokaichi, the Fourth Air Battalion at Tachiarai, the Fifth at Tachikawa and the Sixth Air Battalion at Pyongyang. Three pilot training schools were also established at Tokorozawa, Shimoshizu and Akeno. Initially the Battalions were equipped with a mixture of Sopwith Pups, Nieuport 24Cs and SPAD XIIIs, but with the development of Japanese aircraft they were slowly replaced.

The end of the First World War brought a number of different Allied aircraft to Japan and among them was the Salmson 2A2, which had impressed the Japanese military leaders. The Navy then introduced an expansion programme in 1920, which gave them three squadrons. By 1922 the number had been raised to nine and one year later the Navy put forward a proposal to increase the number of squadrons to seventeen, which would comprise of nine squadrons of seaplanes, six squadrons of landplanes and two training squadrons. This would give the Navy a total of 284 aircraft. But there were a number of high-ranking officials who argued that the heavily armoured and multi-gunned battleships and cruisers were the ultimate weapon in any war at sea, and that wasting money on aircraft that had, in their opinion, no place in the Navy, was pointless. In addition to this, the Washington Disarmament Conference Treaty of 1922 resulted in a massive reduction of finances for the Navy, which cut their expansion programme drastically. Then in 1923 a massive earthquake struck the eastern part of Japan, causing tremendous damage to life and property in Tokyo and the port of Yokohama. This, in turn, caused massive amounts of money destined for the military expansion programme to be diverted for more humanitarian purposes. Despite this, the Navy still managed to increase its Air Arm to ten squadrons by the end of 1923, followed by another three the following year.

The Kawasaki company approached the Army with an offer to build the Salmson 2A2 under licence, and in anticipation of this, had sent a number of the senior engineers from their automobile section to France to study the manufacturing of the aircraft. In the meantime the Army had wanted to manufacture the aircraft themselves, but only had a licence to build the engines. However, they went ahead with production of the aircraft under the pretence of repairing existing Salmson aircraft. The Salmson company in France immediately launched a protest when it discovered what was happening. The manager of the Kawasaki aeroplane section managed to placate Salmson and arranged a licence for Kawasaki to build the aircraft.

In 1921, with the co-operation of the Army, the first of a total of 600 Kawasaki Army Type Otsu 1 Reconnaissance Aircraft was built. They were almost identical to the Salmson 2A2, with just some minor differences around the engine cowling. The first fifty were fitted with 230-hp Salmson engines, but the remainder were powered by a 260-hp Kawasaki Z9 9-cylinder, water-cooled radial engine.

This was one of the first Japanese-built aircraft to see combat, when they saw action in the Manchurian and Shanghai incidents during the Sino-Japanese War. They were initially used in a reconnaissance role but soon became general workhorses, being used as bombers, transport and cargo carriers. They stayed in service until 1933, when the Type 88 replaced them.

The Japanese Army continued to import Western aircraft for training purposes and, if acceptable, the possibility of obtaining licences to build copies of them. One of the aircraft was the French Nieuport 81-E2, and a total of forty of these aircraft were purchased in January 1919. As they were diminishing through wear and tear, or because of crashes or accidents, the Army applied for a licence to build Nieuports as their replacements.

Because the Army did not have the facilities to build aircraft, they handed the contract, drawing and specifications to Mitsubishi. The first of the prototypes came off the production line in May 1922 and was given the designation Army Type Ko 1 Trainer. Two further types, the Ko 2 and Ko 3, came off the production line during the next three years and a total of fifty-seven of the aircraft were built.

THE FIRST JAPANESE CARRIER AIRCRAFT

The increased Navy squadron numbers coincided with the building of the world's first purpose-built aircraft carrier, the *Hosho*, which was capable of carrying nineteen aircraft. As the *Hosho* was nearing completion, the Navy approached the Mitsubishi company with a proposal to build three different types of aircraft: a fighter, a torpedo-bomber and a reconnaissance biplane. Working for Mitsubishi at the time was Herbert Smith from Sopwith, and it was he who designed all three of the aircraft. The fighter, the Mitsubishi 1MF1, was a single-seat biplane of wooden, fabric-covered construction and was powered by a 300-hp Mitsubishi Type Hi 8-cylinder water-cooled engine.

The prototype was produced in October 1921 and delivered to the Provisional Naval Aeronautics Institute (PNAI) for evaluation and flight-testing. After extensive testing the fighter was accepted and became Japan's first naval fighter, given the designation of Navy Type 10 Carrier Fighter (1MF1). There were a number of variations built (1MF1-1MF5), but they only had minor modifications.

The first flights from the aircraft carrier were made by former Flight Lieutenant William Jordan of the RNAS, who was working for the Mitsubishi factory as a test pilot. He made nine take-offs and landings from the deck of the *Hosho* in December 1923.

With these tests completed, the aircraft carrier started operational flights and Lieutenant Shunichi Kira IJNS carried these out in the Type 10 fighter, which had been designed specifically for use on an aircraft carrier.

The second of the aircraft requested by the Navy was the 2MR1 Navy Type 10 Carrier Reconnaissance. The success of the design of the Navy Type 10 Fighter convinced Herbert Smith that an enlarged version of that aircraft would be suitable for a reconnaissance model. On completion, the prototype was test-flown by William Jordan from the Nagoya factory airfield. The aircraft was then sent to the PNAI for test and evaluation and on completion was accepted and an order placed for 128 to be built. Like the fighter, there were four different models produced (2MR1 – 2MR4), the majority of which had only minor modifications. One version however, produced in 1928, was called the Karigane-type, and was modified by moving the radiators from under the fuselage to beneath the wings and having a taller, vertical tail. Although the performance improved, neither the Army nor the Navy liked the version and only the one modified type was built.

A new model of the Type 10 appeared: the 2MRT1, a two-seat training version. Very similar in appearance to 2MR1, the trainer was fitted with dual controls and had a horizontal tail. Again, like previous models, there were variations (2MRT1 – 2MRT3A).

The second version's modification was the positioning of the pilot's cockpit further back down the fuselage.

The Type 10 trainer remained in service with the Navy until it was replaced in 1933 by the Kusho Type 93 Intermediate Trainer.

The third aircraft requested by the Navy was for a torpedo-bomber: the Navy Type 10 Carrier Torpedo Bomber (1MT1N). This was to be the first and only triplane built in Japan, based on similar designs by Sopwith, Fokker and Caproni. The 1MT1N was a two-bay triplane of equal span powered by 450-hp Napier engine, of wooden construction with fabric-covered fuselage and wings. Despite being designed for use on board an aircraft carrier, the wings were non-folding. With the success of the previous two types of aircraft still fresh in their minds, the 1MT1N failed to meet the requirements of the Navy. Once in the air, pilots liked its handling qualities and performance, but on the ground it was extremely difficult to handle because of its 14-foot height. Production started and twenty of the aircraft were built, but production was halted with the introduction of a new Type 13 model that superseded it.

In 1921 the British Aviation Mission was assisting the JNAF in the training of its pilots, and to aid them ten types of British aircraft were sent to Japan aboard a freighter. One of Britain's best flying boats was amongst them, the Felixstowe F5, built by the Short Brothers. The original intention was to build this aircraft in Japan, and in anticipation of this twenty-one engineers were sent there.

Construction work started in April 1921 at the Yokosuka Arsenal and a number of Japanese naval personnel were placed under the leadership of Short Brothers' chief engineer, Charles Dodds. In addition to licensing the Japanese to build the F5, they also supplied partially-built assemblies, together with tooling and manufacturing instructions, so that the first six could be constructed easily. The first model came off the production line in less than a month and the first test flights excelled all expectations. By the time the last of the six came off the production line, the Japanese were ready to build their own versions.

The contract to build the aircraft was given to the Aichi company, who between 1921 and 1929 built forty F5 flying boats. The engines used in the first six F5s were the 360-hp Rolls-Royce Eagles, but as work developed at the engine factory of the Hiro Arsenal, they produced a 400-hp Lorraine engine that was fitted into the F5. The Japanese hoped that this would be the engine to power their version of the F5, but it was never as reliable as the Rolls-Royce engine.

The Navy F5 Flying Boat entered service with the Navy at the end of 1921 and was used as a long-range reconnaissance/patrol aircraft until 1930, carrying out numerous long-distance flights. There were a number of problems with the Japanese-built F5, mainly due to engine faults and poor maintenance. A total of sixty F5s were built and all saw service with the Navy.

At the beginning of the 1920s the Imperial Japanese Navy was in the process of a major shipbuilding programme called the 8-8 Fleet Plan. This was to consist of eight battleships, fitted with 18-inch guns, and eight cruisers. However, while the work was being carried out, the Washington Disarmament Conference Treaty was signed, so two of the cruisers were converted into aircraft carriers with a displacement of 34,000 tons. They were called the *Akagi*, which was completed in 1927, and the *Kaga*, which was completed in 1928.

The need for a seaplane training aircraft prompted the design of the Yokosuka Yi-Ko Floatplane, which was a complete success and after a series of flight tests it was put into production. A number of different engines were tried, including the 200-hp Hispano-Suiza, the 100-, 110- and 130-hp Benz and the 70- and 100-hp Renault. This was the Japanese Navy's first real seaplane trainer and seventy of the aircraft were built at the Yokosuka arsenal between 1920 and 1922. The first twenty-four were fitted with either the 130-hp Benz (ten) or the 70-hp Renault (ten), 100-hp Renault (two) and 100-hp Benz. The next forty-two were fitted with 110-hp Benz (thirty-six) and 200-hp Hispano-Suiza (six) followed by 130-hp Benz (four).

The two principal aircraft used by the Navy during this period were the Hansa-Brandenburg 310 twin-float reconnaissance seaplane and the Avro 504K and L models. Over 300 Hansa-Brandenburgs were built, 160 by Nakajima and 150 by Aichi, and 250 Avro 504K and L models were built, 150 by Nakajima and 100 by Aichi.

The development of the seaplane prompted the converting of a captured Russian freighter, the *Lethington*, to a seaplane carrier renamed the *Wakamiya* in 1914. The *Lethington* had been built as a Russian freighter ship in Glasgow in 1900 and had been captured in 1905 by a Japanese torpedo boat while on a voyage from Cardiff to Vladivostok during the Russo-Japanese War. Renamed the *Wakamiya-Maru*, it was used initially as a transport ship until 1913, when it was taken over by the Japanese Navy and converted into a seaplane carrier.

When converted in August 1914, the 7,720-ton *Wakamiya* carried four aircraft, which were launched by means of a crane that lowered and recovered the aircraft after their missions. In September 1914, her aircraft carried out the first naval-launched air raids when German-held land targets were bombed. The following month, during the Siege of Tsingtao, raids were carried out on ships anchored in Qiaozhou Bay.

She remained in service until 1918, when it was replaced by a converted oil tanker, the *Notoro*, which had the capability of carrying eight seaplanes. This also heralded the development of a launching system, which enabled aircraft to be launched from a variety of ships, including battleships and cruisers. In April 1920 the *Wakamiya* was converted once again; this time she had a launch platform on the foredeck. In June the same year, the first take-off of a Japanese aircraft from an aircraft carrier was achieved, paving the way for the development of Japan's first purpose-built aircraft carrier, *Hosho*. The *Wakamiya* was used for trials of aircraft launching until April 1931, when she was scrapped.

With the development of the purpose-built aircraft carrier, the *Notoro* was converted back to an oil tanker and stayed that way until scrapped in the 1930s.

RIVALRY, AND A GROWING AVIATION INDUSTRY

It became very apparent that as the military increased its strength and capability, so the rivalry between the Army and the Navy would become increasingly bitter. The Government, in an attempt to defuse the situation, created a joint Army/Navy Co-ordination Committee. The aim of the committee was to find a peaceful solution to the rivalry that was threatening to create an unstable defence system at a time of growing unrest. It was proposed that an independent air force be created from the air arms of both the Army and the Navy, but the Navy, ever fearful of the Army's attempt to dominate, rejected the idea. In fact neither service wanted to lose their air arms and so the talks collapsed.

In the meantime, some of the major companies saw the opportunity to create an aviation industry. The Tokyo Gas and Electric Co. began building airframes and engines, becoming the Hitachi Aircraft Company. Kawasaki Shipbuilding Co. created a separate company, Kawasaki Aircraft. Kawanishi Machinery Co. became Kawanishi Aircraft. And the Aichi Clock and Electric Company became Aichi Aircraft. At the beginning of 1920, the Navy built a new arsenal at Hiro and started producing its own aircraft.

One of the existing arsenals, the Yokosho Arsenal, was tasked at the beginning of 1924 by the Navy with providing a replacement for the I-go Ko-gata Trainer; only this time they wanted an aircraft that had an interchangeable wheel and twin-float undercarriage. The specifications required a single-engine, twin-float seaplane/land-based open cockpit trainer. One year later the first prototype, of wooden construction with fabric covered wings, tail section and fuselage, emerged from the arsenal and was immediately sent for test and evaluation. With all the tests completed the Navy accepted it, and production orders were given to the Kawanishi, Nakajima and Watanabe companies.

Over the next ten years the three companies produced over 100 of the Yokosho Navy Type 13 Trainer (K1Y2) seaplanes. The landplane versions were designated the Navy Type

13 Trainer (K1Y1) and were used right up to the beginning of the Pacific War. Such was the demand for the aircraft that three more companies were tasked with manufacturing it: Kawanishi, Nakajima and Watanabe. Between them they manufactured 104 Navy Type 13s.

One of the most successful carrier aircraft built at the time was the Navy Type 13 Carrier Attack Aircraft (B1M1). Designed by Herbert Smith of Sopwith, it had been born out of the need for a carrier attack bomber requested by the JNAF. It was a single-engined, two-seat, three-bay biplane constructed entirely of wood with fabric covering. It had a wide-track fixed undercarriage and wings that folded back for ease of stowage aboard aircraft carriers.

Powered by a 450-hp Napier Lion 12-cylinder, water-cooled engine, the aircraft was sent to the Navy in 1923 for test and evaluation after successfully completing company tests. The B1M1 was accepted by the Navy and put into production the following year, using the Napier Lion engine. The Type B1M2 appeared the following year, this time powered by a 450-hp Mitsubishi Type Hi 12-cylinder Vee water-cooled engine. The aircraft went into production as a three-seater biplane.

The JAAF were also very impressed with the B1M1, so much so that it was developed as the Army Type 87 Light Bomber. With the B1M2 now accepted by the Navy, modifications were carried out on this version to improve the take-off and climb performance. Installing the Farman reduction gear in the engine and fitting a large-diameter four-bladed propeller achieved this. Accepted by the Navy as the Type 13-3 (B1M3), it became the Navy's main all-round combat aircraft until the beginning of the Sino-Japanese conflict. It was involved in one of the first air-to-air combat engagements of the war when, together with six other naval fighter aircraft, it shot down a Chinese Boeing P-12 fighter. During the engagement two of the Japanese aircraft were shot down. Over 650 B1M1/2/&3s were built between 1923 and 1930 and they were highly regarded by Western aircraft manufacturers and military alike.

The reason for the sudden upsurge in demand was that companies realised that the development of aviation offered them a piece of a very lucrative market as the needs of the military grew. Such was the influence that the military had wealthy industrialists queuing up to invest money in aircraft production. Up to this point almost all aircraft were European models that were being built under licence, and so it came as no surprise when the major Japanese companies started to develop close ties with European manufacturers. Kawasaki associated itself with Dornier, whilst Mitsubishi acquired the help of Junkers and Rohrbach. They also employed Herbert Smith of Sopwith and Dr Alexander Baumann as engineering instructors. Nakajima allied itself with Breguet and Nieuport, and slowly the whole Japanese aviation industry started to take shape.

A NEW SEAPLANE

In an effort to replace the catapult-launched reconnaissance seaplanes of the Hansa Type, the JNAF put out a requirement in the shape of a competition for a short-range seaplane capable of being launched from cruisers and battleships. In 1926 three companies were approached, Aichi, Nakajima and Yokosho, all with experience of manufacturing seaplanes. The Aichi company submitted a proposal for a low-wing monoplane based on the proven design of the Hansa Type, but with the modifications in place to eliminate the known faults. The Yokosho proposal was for a low-wing monoplane similar to that of the Aichi, but constructed of metal with the tailplane mounted on top of the fuselage and the rudder protruding way below the level of the fuselage. The rectangular-section fuselage was of the cantilever type and mounted on twin floats by four struts. This model was rejected because of its in-flight instability, but it became the first all-metal aircraft to be built in Japan.

The Nakijima proposal was for a sesquiplane with a good downward view for the two crewmembers. All three proposals had been fitted with the 300-hp Mitsubishi Type Hi engine. The Nakajima entry was the one selected and it became Japan's first originally-designed shipboard reconnaissance seaplane.

There were a number of initial teething problems and it was not until May 1927 that the Navy accepted the aircraft, giving it the designation Navy Type 15 Reconnaissance Seaplane (E2N1). A later model, the E2N2, was fitted with dual controls and a hood for instrument training. The E2N1 was a single-engined twin-float seaplane of an all-wooden, fabric-covered construction and armed with a single flexible 7.7 mm machine gun mounted in a dorsal position. It was only used for short-range reconnaissance missions.

Nakajima produced the aircraft for the first year (1928-1929) then manufacture was transferred to Kawanishi. Earlier models of this type of aircraft used to launch from cruisers and battleships, from the tops of gun turrets, using their own power, but the E2N1 was the first to use the new power catapult system and could be launched from almost any position. Between 1927 and 1929 the Nakajima company built forty-seven of the aircraft, the Kawanishi company built thirty between 1929 and 1930. The Nakajima Type 90 eventually replaced the E2N1.

In February 1928 the JNAF put out a requirement for a replacement for the Type 13 Carrier Attack Aircraft, only this one was to be a carrier attack bomber. Four of the major companies, Aichi, Kawanishi, Mitsubishi and Nakajima, put forward proposals. The requirements were that it was to have a crew of three, made of part metal and part wood, have an endurance of at least three hours with a bomb load (or more than eight hours without), and an operating ceiling of over 19,686 feet (6,000 m). It had to have a maximum speed of over 126 mph (110 knots) and a deck landing speed of 51 mph (45 knots). The specifications also laid down the measurements of the aircraft: the wingspan had to be less than 49 ft 2 in. (15 m), the length less than 32 ft 8 in. (10 m) and the height less than 12 ft 4½ in. (3.8 m).

In March 1928, the Mitsubishi company created three different design studies. The first, known as 3MR3, was subcontracted to the Sopwith designer/engineer Herbert Smith, who had just returned to England. A second study, 3MR4, was given to the Blackburn company in Britain, whilst the third, 3MR5, was given to the Handley Page company, also in Britain. As this shows, the Japanese aviation industry was still heavily reliant on the Western world for its technical expertise and designs.

All three companies submitted their designs at the end of the year, and Mitsubishi selected the 3MR4 and asked the Blackburn company to make the first prototype powered by a 600-hp Hispano Suiza engine. At the same time Mitsubishi sent some of the engineers to the Blackburn company to study engineering planning, design and to watch the prototype being built and to learn from it. The aircraft built was a single-engined, two-bay biplane made of welded steel tubing covered in aluminium and fabric. The wings were constructed in the same way and covered in fabric, and were rearward folding for ease of stowage aboard aircraft carriers.

The first flight of the prototype took place in England, and when satisfied with the results, the company shipped the aircraft to Japan in February 1930. With the aircraft came the Blackburn chief designer, who supervised the assembly of the aircraft. A second prototype, powered by a 650-hp Mitsubishi Type Hi engine, followed in the October, while the first one was still undergoing test and evaluation. The second prototype was lost when it crashed due to pilot error, so a third prototype was ordered which arrived in February 1931. But problems arose with engine overheating and controllability when it was making a three-point landing, a manoeuvre essential for carrier landings.

A fourth prototype was built with all the shortcomings and problems of the previous three resolved. After being tested by the Japanese Navy, that aircraft was accepted and put into production with the official designation Mitsubishi Type 89-1 Carrier Attack Aircraft (B2M1).

Within months of coming off the production line, problems arose with the engine and a number of other minor things. Production was stopped while the problems were resolved, and the modified aircraft came off the production line with the designation Type 89-2 (B2M2).

The aircraft first saw action during the Shanghai incident and took part in a number of missions. It earned the approval of the pilots who flew it and was operational until 1935. A total of 205 B2M1 and M2s were built.

NEW ARMY BOMBERS

Both the Army and the Navy took a long hard look at their air arms and realised that it was essential that a strong domestic aircraft industry be developed. All their aircraft were either imported foreign aircraft or aircraft built in Japan under licence from foreign companies. The competition had produced an aircraft that had a strong Japanese input and it was this that gave the military the impetus to push for Japanese-designed and -built aircraft.

The Army's need for a heavy bomber prompted them to approach the Kawasaki company to build an all-metal aircraft to replace their two ageing heavy bombers, the Farman F.50 and F.60 Goliath. Realising that they still didn't have the necessary expertise to build such an aircraft, they approached Dornier and BMW for help. In March 1924, Tomokichi Takezaki and a team of engineers went to the Dornier factory in Switzerland to obtain the licence to build one of their aircraft in Japan.

With the licence approved, the team returned to Japan together with a team of Dornier engineers. The aircraft had already been designed in Switzerland and, under orders from the Army, the whole project was to be kept top secret. The first prototype was completed at the beginning of 1926, but a 450-hp Napier Lion engine that had been imported from Britain powered them. This was because the BMW IV engine, being built by Kawasaki, was not ready. A second prototype appeared in April and was powered by the BMW IV engine.

Both prototypes were put through their paces by Army test pilots over the following year, performing better than expected. In 1927 they were put into production without any modifications being necessary. A further twenty-six of the aircraft were built and assigned to bomber squadrons and briefly saw action in Manchuria during the Sino-Japanese War.

The Army Type 87 Heavy Bomber carried a crew of six: pilot, co-pilot, navigator, radio operator, engineer/gunner and a bombardier/nose gunner. They were not that popular with the Army crews because with a one-ton bomb load they were ponderous and very heavy to handle, but at that moment in time they were the best available.

The need for a new reconnaissance aircraft was still high on the Army's agenda, and in 1926 they hosted a competition with the Kawasaki, Mitsubishi, Ishikawajima and Nakajima companies to design an all-metal long-range reconnaissance aircraft. All the companies looked towards German designers to help them in their projects when the Army laid down the specifications. They wanted a maximum speed above 124 mph (200 km/h), a range of over 600 miles (1,000 km), a capability of carrying one forward-firing fixed 7.7 mm machine gun, twin 7.7 mm flexible machine guns mounted in the dorsal position, an aerial camera and a radio.

It was Kawasaki who produced the first single-engined biplane, powered by BMW V1 engines and built by Kawasaki under licence. It had a stressed metal skin on the forward section of the fuselage, the remainder being fabric-covered. During the initial tests the KDA-2, as the company called it, was found to have exceeded the Army's specifications. The Army ordered a further two prototypes to be built and additional equipment to be installed. Both aircraft still exceeded the specification laid down. The other companies by this time had withdrawn from the competition, and the 200,000-yen prize was awarded to the Kawasaki company.

The prototypes were sent to the Army's test facility at Tokorozawa, during which time, after some modifications, they were used in successful air-to-air refuelling tests. They were also fitted with a Sperry autopilot and were involved in tests with a three-bladed propeller. One of the prototypes was converted to a seaplane but was not successful.

Accepted by the Army, the Army Type 88 Reconnaissance Aircraft, as it became known, went into production. The first models, which were known as the Type 88-1s, had had a flat frontal radiator. In a demonstration of their long-range capability, two of these models were fitted with additional fuel tanks and made an over-water flight from Tachiarai in Kyushu to Pingtung in Taiwan, a total of 750 miles in eight hours. A remarkable flight for the time.

Production models were now being sent to front-line units and they saw action in Manchuria and Shanghai during the Sino-Japanese War. Later models were known as the Type 88-2s, the only real difference being that they had a tapered nose with a spinner attached to the propeller and a taller, tapered fin and rudder.

A total of 710 Army Type 88 Reconnaissance aircraft were built between 1927 and 1931.

At the same time as the Type 88 made its appearance, the Army still had a need for a light bomber, but one that was of all-metal construction. The use of wood and fabric in constructing aircraft had reached the limit to which it could be developed successfully, so it was decided to use the Kawasaki Type 88-2 model reconnaissance aircraft and develop it as a bomber. One of the Type 88-2s was taken off the production line, given a strengthened lower wing and fitted with bomb racks beneath the fuselage, together with a bombsight.

The aircraft was tested by the Army in its new guise, and was immediately accepted and placed into production. Such was the demand for the new bomber that the Tachikawa company was asked to build it and produced thirty-seven during the next three years. In total, 407 of the aircraft were built between June 1927 and December 1933.

But still the aircraft being built were all of European design, mainly because they were the tried and tested models. Then in 1925 the Army put out a tender for a light bomber, and invited manufacturers to enter into competition for a Japanese-designed and -constructed aircraft. Three companies entered the competition, Kawasaki, Mitsubishi and Nakajima. But still the companies could not shake off the European influence, as the Kawasaki was based on the Dornier Komet. Mitsubishi entered two aircraft, the 2MB2 and 2MB1; the latter was also known as the Type 87. Nobushiro Nakata and Satsuo Tokunaga designed the 2MB1 jointly under the supervision of Dr Alexander Baumann. Known as the Washi (Eagle)-type light bomber, it was regarded as the best of the all entries but was deemed unacceptable because of its very high unit cost. The company immediately put forward the 2MB1, a highly modified version of the Navy Type 13 carrier attack aircraft. Despite being regarded as obsolete, it was accepted by the Army because of its excellent handling and its relatively low unit cost.

The Type 87 2MB1 was fitted with a 450-hp Hispano-Suiza engine and the radiator fitted in front of the engine instead of along the sides as in the Type 13. The aircraft was also given dual controls and changed into a two-seater. The folding-wing capability was removed, as was the dihedral on the upper wing.

After extensive testing the Mitsubishi Army Type 87 Light Bomber (2MB1) was put into production, and when fighting erupted in Manchuria the bomber was involved in the early part, supporting ground troops. Then production was suspended when the Army took delivery of the Kawasaki Type 88 Light Bomber and Reconnaissance aircraft. The remaining 2MB1s were used for reconnaissance and training and a total of forty-eight of the aircraft were built between 1926 and 1928.

NEW FLYING BOATS

Whilst the Army were slowly increasing their air arm, the JNAF was also developing new flying boats based on western designs. In 1926 it decided that a replacement for the Japanese-built Felixstowe F5 was needed. The F5 had entered service in 1921 and had proved itself to be a most reliable aircraft, especially over long distances. The wooden construction of the new flying boat was based on the proven hull of the F5, as were the floats, but unlike the F5, the wings were of a single-bay biplane design with metal wingtip floats. Two 400-hp Lorraine 1 engines built by the Hiro company powered the aircraft. Given the designation Hiro Navy Type 15 Flying Boat (H1H1), the prototype was sent to the Navy for test and evaluation, and although some minor modifications were recommended, it was accepted and production orders were made.

The first of three prototypes off the production line were powered by the 400-hp Lorraine 1 engines, but the production models, Type 15-1 (H1H1), were powered by 450-hp Lorraine engines. The first models had bench-type aileron balance surfaces, but subsequent models had horn-balanced ailerons fitted. In an effort to improve stability, additional vertical ailerons were fitted close to the tips of the tailplane.

The second of the prototypes, the H1H2, was an experimental project that had an all-metal hull with increased horn-balance aileron tabs fitted to the wings. While experiments were going on with the H1H2, the Yokosuka Arsenal was developing another prototype, the H1H3. This aircraft had an all-metal hull and wingtip floats, and equal span wings made of wood and covered in fabric. It was powered by two 450-hp Lorraine 3, 12-cylinder, water-cooled engines that turned four-bladed wooden propellers.

The H1H3 actually came out at the same time as the H1H1, but because of a delay in the new engines being accepted, it was two years before they went into service. The H1H1/3 flying boats remained in service until 1938, proving that the Japanese could design and build flying boats of quality without relying on the Western world to provide them with inspiration and technology. The Hiro, Yokosho and Aichi companies built a total of sixty-five H1H1/3s between 1927 and 1934.

As the H1H1/3s came off the production line in 1928, the Yokosuka company was already looking for a replacement flying boat. This was to be based on the design of the Supermarine Southampton biplane flying boat that had been imported from Britain in 1927. After extensive testing, the Southampton was flown to the Hiro Arsenal so that the designers there could make a close inspection of the aircraft. The prototype appeared in 1930 and appeared at first glance to be almost identical to the Supermarine Southampton. There was a noticeable difference, however; the triple fin and rudders of the Southampton had been replaced with a single fin and rudder. The hull was of an all-metal semi-monocoque construction with the upper surface of the hull rounded to meet the bottom section of the hull, thus dispensing with the flared chines of the earlier designed hulls. The wing structures were also made of metal but covered in fabric.

After initial testing the aircraft, designated the Navy Type 89 Flying Boat (H2H1), was tested by the Navy and accepted. The second prototype was also tested some months later, but an incident when a fuel line ruptured, causing the aircraft to catch fire, put the programme back. The pilot managed to land close to the beach, where the crew abandoned the aircraft to the flames. Despite this setback, the Navy placed contracts for the aircraft, and such was the demand that the Aichi and Kawanishi companies were also involved in the production.

The H2H1 saw action during the Sino-Japanese conflict and was the last of the Japanese twin-engined biplane flying boats. A total of seventeen of the aircraft were built between 1930 and 1931.

Towards the end of the 1920s the design and development of aviation in Japan had slowed noticeably. The facilities had grown and the standard of workmanship improved, but the military hierarchy was reluctant to let go of the samurai warrior legacy that had been the cornerstone of its military existence for centuries. Because of their geographical position, Japan maintained that the biggest threat to it came from the Republic of China, the Soviet Union and the United States. It was considered that the Army could deal with any threat coming from either China or the Soviet Union, while the Navy could deal with the United States. They saw the role of the Army Air Arm as nothing more than support for the ground forces, and the Navy Air Arm as a reconnaissance tool.

New aircraft were slowly coming from the various companies, but the majority of them were either copies of European and American aircraft or designed by Europeans. The first of two design competitions to find a new torpedo bomber for the Navy to replace the Type 13 took place in 1928. The four major companies, Mitsubishi, Kawanishi, Aichi and Nakajima, entered the competition. The only one produced that met the Navy's specifications was the Mitsubishi 3MR4, also known as the B2M1, a three-seat biplane powered by a 600-hp Hispano-Suiza engine. Once again a European company, the Blackburn Aircraft Company, built the aircraft at the request of Mitsubishi and it was designated the Type 89. Over 200 of the aircraft were built.

A SUBMARINE-LAUNCHED SEAPLANE

In 1923, the Japanese Navy had purchased two Heinkel-Caspar U-1 aircraft with the idea of using them in the same way as the German Navy had earlier, but it was 1927 before trials had been completed. By this time the Japanese had developed their own aircraft, the Yokosho 1-go, and a method of transporting it on a submarine. The Yokosho 1-go was almost identical to the U-1, but with one or two modifications and a more powerful rotary engine. It was of a cantilever construction, with the upper and lower fabric-covered wings having no interplane struts. The twin-floats extended beyond the nose of the aircraft and were designed, like the wings, to be detached quickly and easily. The light metal fuselage could then be slid into a tube-like hangar fixed to the deck of the submarine, together with the wings and the floats. Five mechanics could assemble the aircraft in four minutes and it could be airborne in fifteen minutes. Dismantling the aircraft once it was craned back aboard took two minutes, and the submarine could start submerging in less than ten minutes.

The first test flight of the Yokosho 1-go was carried out aboard the mine-laying submarine I-121 in. 1925, and despite a number of problems was deemed to have been successful. The aircraft was operated from the submarine I-5 for eighteen months, but the submarine was too slow and the aircraft was transferred to the I-51 Class submarines in 1930. A number of new innovations were made during the initial period, with the result that the I-51 had a compressed air catapult fitted to her afterdeck, together with a hangar capable of taking two aircraft.

With the development of aircraft gathering pace, the need for more pilots became a priority, and therefore more training aircraft. Early in 1928 it was obvious that the Avro 504 that had been used as a trainer was obsolete, and so the Navy ordered the Yokosuka arsenal to design a new land-based trainer. What in fact emerged was a modernised version of the Avro 504, only this time it was powered by a 130-hp Armstrong Siddeley Mongoose radial engine in place of the 110-hp Le Rhône rotary engine. Like the Avro 504, it was of an all-wood construction covered in fabric with open cockpits in tandem. The undercarriage and tail section were re-designed and after extensive testing the aircraft, given the designation Yokosuka Type 3 (K2Y1), was accepted by the Navy as a trainer. The Yokosuka arsenal built the first six and further production was put out to other companies, Kawanishi (sixty-six), Watanabe (114), Mitsubishi (forty-five), Nippi (126) and Showa (three). In total 360 of the aircraft were built over the next ten years and it became the primary trainer for the Navy.

It was a well-respected training aircraft because of its ease of handling and manoeuvrability, and the Manchurian National Military Force acquired a small number.

FURTHER NAVAL EXPANSION AND REORGANISATION

In 1928, the Navy established an aviation bureau known as the Koku-Hombu, giving it overall control of naval aviation, which became known as the Imperial Japanese Navy Air Force (IJNAF). The bureau immediately got to work developing a building programme and establishing bases for the landplanes. They acquired six land bases at Yokosuka, Kure,

Sasebo, Ohmura, Kasumigaura and Tateyama. There were additional seaplane and flying boat squadrons.

As the development of aircraft progressed, the Navy decided to improve its reconnaissance fleet of aircraft by issuing a requirement to Aichi, Nakajima and Kawanishi for a three-seat reconnaissance seaplane to replace the Type 14 Reconnaissance Seaplane with one using the latest metal technology. The Yokosuka Arsenal had not been approached, but had decided to start its own project and once again referred to the Western world for guidance by acquiring a Heinkel HD 28 seaplane.

The first of the Yokosuka prototypes appeared in August 1928 and initially was given the designation Type 14-2 Reconnaissance Seaplane. It was a single-engined, twin-float seaplane with a welded steel fuselage, which was stepped down on the lower section to allow for a ventral downward-firing machine gun covered in fabric. The wings were of fabric-covered wooden construction and had a slight stagger. The twin floats were of an all-metal construction and were fixed to the fuselage by two inverted Vee struts. It was powered by a 450-hp Nakajima Jupiter VIII radial engine, which gave it a top speed of 110 mph (70 knts).

Two pre-production models were ordered and built by Kawanishi in October 1931. Both aircraft were then evaluated by the Navy, and after a number of minor modifications it was accepted by the Navy and given the designation Navy Type 90-3 Reconnaissance Seaplane (E5K/Y1).

After the first five had been built, it was noticed that the overall performance of the older Type 14 reconnaissance seaplane was better, so it was decided to install the 500-hp Nakajima Jupiter IX engine into the next seventeen, but the difference was negligible. Their arrival came when the Shanghai Incident flared up, and the E5Y1s were assigned to battleships and seaplane tenders and almost all saw service during the conflict.

Only twenty of the aircraft were ever built, but it heralded the start of a new phase in aircraft manufacturing in Japan.

In April 1926, the Navy had put out a request for a new carrier fighter to replace their aging Mitsubishi Type 10 Carrier Fighter. They asked the three major companies, Aichi, Nakajima and Mitsubishi, to submit proposals. Nakajima was well aware of the strides British aircraft companies had made and contacted the Gloucestershire Aircraft Company to build them a modified version of their very successful Gamecock Gamber.

It was around this time that the aircraft carriers *Akagi* and *Kaga* were commissioned; both were capable of carrying sixty aircraft each. Initially the range of the carrier-based aircraft was limited to a 100-mile radius, but as the Navy developed, so they demanded aircraft with longer and longer-range capabilities. The realisation that the aircraft carrier could play a major part in any war also made them aware of the need for a long-range reconnaissance flying boat. They had been considering the use of the airship, but the tragic loss of their N.3 airship and the subsequent loss of life put the use of this type of craft for long-range deployment in doubt.

Despite the restrictions of the Washington Treaty both the Army and the Navy increased their Air Wings, and by 1929 the Army, now known as the Imperial Japanese Army Air Force (IJAAF), had eight Air Wings, the majority of which were made up of fighter and reconnaissance squadrons. The only bomber wing, the Seventh Air Wing, was made up of four squadrons based at Hamamatsu and attached to the Third Ground Force Division. The role of the bomber wing was seen as purely a support for ground troops and no provision was made for long-range bombing to be developed. This was to be a major cause for concern for the Japanese during the Second World War, and to have far-reaching consequences.

SPECIFICATIONS

KAWASAKI TYPE OTSU 1 RECONNAISSANCE AIRCRAFT

Wing Span:	38 ft 7¼ in. (11.8 m)
Length:	28 ft 3 in. (8.6 m)
Height:	9 ft 6 in. (2.90 m)
Weight Empty:	2,050 lb (930 kg)
Weight Loaded:	3,306 lb (1,500 kg)
Max. Speed:	116 mph (101 kts)
Ceiling:	19,028 ft (5,800 m)
Endurance:	7 hours
Engine:	230-260-hp Kawasaki-Salmson Z.9, 9-cylinder water-cooled, radial
Armament:	One fixed forward-firing 7.7 mm machine gun and one dorsal 7.7 mm machine gun.

MITSUBISHI ARMY TYPE KO1 TRAINER

Upper Wing Span:	33 ft 8 in. (10.26 m)
Lower Wing Span:	33 ft 8 in. (10.26 m)
Length:	22 ft 4¾ in. (7.13 m)
Height:	10 ft (3.05 m)
Weight Empty:	1,212 lb (550 kg)
Weight Loaded:	1,763 lb (800 kg)
Max. Speed:	72 mph (63 kts)
Ceiling:	13,123 ft (4,000 m)
Endurance:	1 hour
Engine:	100-hp Le Rhône, 9-cylinder, air-cooled rotary
Armament:	None

NAKAJIMA ARMY TYPE KO 2 TRAINER

Upper Wing Span:	26 ft 7¼ in. (8.11 m)
Lower Wing Span:	26 ft 7¼ in. (8.11 m).
Length:	23 ft 1 in. (7.03 m)
Height:	9 ft 6 in. (2.9 m)
Weight Empty:	970 lb (440 kg)
Weight Loaded:	1,565 lb (710 kg)
Max. Speed:	87.5 mph (76 kts)
Ceiling:	16,404 ft (5,000 m)
Endurance:	2 hours
Engine:	80-100-hp Le Rhône, 9-cylinder, air-cooled rotary
Armament:	None

NAKAJIMA ARMY TYPE KO 3 FIGHTER/TRAINER

Upper Wing Span:	26 ft 11½ in. (8.22 m)
Lower Wing Span:	26 ft 11½ in. (8.22 m)
Length:	18 ft 7¼ in. (5.67 m)
Height:	7 ft 10½ in. (2.40 m)
Weight Empty:	915 lb (415 kg)
Weight Loaded:	1,311 lb (630 kg)
Max. Speed:	85 mph (74 kts)
Ceiling:	16,404 ft (5,000 m)

Endurance: 2 hours
Engine: 80-93-hp Le Rhône, 9-cylinder, air-cooled rotary
Armament: One forward-firing 7.7 mm machine gun

Nakajima Army Ko 4 Fighter

Wing Span: 31 ft 9¾ in. (9.7 m)
Length: 21 ft 1½ in. (6.4 m)
Height: 8 ft 8 in. (2.64 m)
Weight Empty: 1,818 lb (825 kg)
Weight Loaded: 2,557 lb (1,160 kg)
Max. Speed: 145 mph (126 kts)
Ceiling: 26,246 ft (8,000 m)
Endurance: 2 hours
Range: Not known
Engine: One 300-320hp Mitsubishi-Hispano Suiza 8-cylinder, water-cooled
Crew: One
Armament: Two fixed forward-firing 7.7 mm machine guns

Hiro Navy Type F5 Flying Boat

Wing Span: 103 ft 8 in. (31.59 m)
Length: 49 ft 4 in. (15.03 m)
Height: 18 ft 10¼ in. (5.75 m)
Weight Empty: 8,201 lb (3,720 kg)
Weight Loaded: 12,405 lb (5,627 kg)
Max. Speed: 102 mph (89 kts)
Ceiling: 11,646 ft (3,550 m)
Endurance: 7 hours
Range: 712 miles (620 nm)
Engine: Two 360-hp Rolls-Royce Eagle, 12-cylinder, water-cooled
Crew: Four
Armament: Two flexible 7.7 mm machine guns

Mitsubishi 2MB1

Wing Span: 48 ft 6¾ in. (14.8 m)
Length: 32 ft 9½ in. (10.00 m)
Height: 11 ft 11 in. (3.63 m)
Weight Empty: 3,968 lb (1,800 kg)
Weight Loaded: 7,275 lb (3,300 kg)
Max. Speed: 115 mph (100 kts)
Ceiling: 14.025 ft (4,275 m)
Endurance: 3 hours
Engine: 600-hp Mitsubishi-Hispano-Suiza, 12-cylinder water-cooled.
Armament: Two fixed forward-firing 7.7 mm machine guns and one dorsal 7.7 mm
                machine gun and one ventral 7.7 mm machine gun

Mitsubishi Experimental Washi-type Light Bomber (2MB2)

Upper Wing Span: 65 ft 7½ in. (20 m)
Lower Wing Span: 44 ft (13.4 m)
Length: 32 ft 3¼ in. (9.85 m)
Height: 13 ft 5½ in. (4.10 m)

Weight Empty: 4,629 lb (23,100 kg)
Weight Loaded: 8,024 lb (3,640 kg)
Max. Speed: 131 mph (114 kts)
Ceiling: 19,685 ft (6,000 m)
Endurance: 3 hours
Engine: 600-hp Mitsubishi-Hispano-Suiza, 12-cylinder, water-cooled
Armament: Two fixed forward-firing 7.7 mm machine guns and one dorsal 7.7 mm
                machine gun and one ventral 7.7 mm machine gun

Yokosho Navy Type 91 Reconnaissance Seaplane (E6Y1)

Wing Span: 26 ft 3 in. (8.0 m)
Length: 21 ft 11½ in. (6.69 m)
Height: 9 ft 5 in. (2.87 m)
Weight Empty: 1,256 lb (570 kg)
Weight Loaded: 1,653 lb (750 kg)
Max. Speed: 104 mph (91 knts)
Ceiling: 9,483 ft (3,000 m)
Endurance: 4.4 hours
Range: Not known
Engine: One 130-150-hp Mongoose 5-cylinder, air-cooled radial
Armament: None

Navy Type 90-3 Reconnaissance Seaplane (E5Y1)

Wing Span: 47 ft 5¼ in. (14.46 m)
Length: 35 ft 5¾ in. (10.8 m)
Height: 15 ft 6¾ in. (4.74 m)
Weight Empty: 4,078 lb (1,850 kg)
Weight Loaded: 6,613 lb (3,000 kg)
Max. Speed: 110 mph (96 knts)
Ceiling: 13,287 ft (4,050 m)
Endurance: 6½ hours
Range: Not known
Engine: One 450-520-hp Nakajima Jupiter VIII 9-cylinder, air-cooled radial
Armament: One dorsal flexible 7.7 mm machine gun, one flexible ventral 7.7 mm
                machine gun and two forward-firing fixed 7.7 mm machine gun. Two
                275 lb (125 kg) or three 132 lb (60 kg) bombs

Hiro Navy Type 15 Flying Boat (H1H1-2)

Wing Span: 75 ft 4 in. (22.9 m) (H1H1)
                72 ft 2 in. (22.0 m) (H1H2)
Length: 49 ft 7 in. (15.11 m) (H1H1)
                52 ft 2½ in. (15.9 m) (H1H2)
Height: 17 ft (5.19 m) (H1H1)
                17 ft 11¼ in. (5.4 m) (H1H2)
Weight Empty: 8,862 lb (4,020 kg) (H1H1)
                9,810 lb (4,450 kg) (H1H2)
Weight Loaded: 13,448 lb (6,100 kg) (H1H1)
                14,330 lb (6,500 kg) (H1H2)
Max. Speed: 106 mph (92 knts) (H1H1)
                104 mph (90.5 knts) (H1H2)
Ceiling: 9,483 ft (3,000 m) (H1H1&2)

Endurance:	14.5 hours
Range:	Not known
Engine:	Two 450-hp Lorraine 2, 12-cylinder, water-cooled
Armament:	One bow-mounted flexible 7.7 mm machine gun and one dorsal-mounted flexible 7.7 mm machine gun

MITSUBISHI B2M1

Wing Span:	49 ft 11¼ in. (15.22 m) (B2M1)
	49 ft 1¾ in. (14.98 m) (B2M2)
Length:	33 ft 8½ in. (10.27 m) (B2M1)
	33 ft 4¾ in. (10.18 m) (B2M2)
Height:	12 ft 2 in. (3.71 m) (B2M1)
	11 ft 9½ in. (3.60 m) (B2M2)
Weight Empty:	4,982 lb (2,260 kg) (B2M1)
	4,806lb (2,180 kg) (B2M2)
Weight Loaded:	7,936 lb (3,600 kg) (B2M1&2)
Max. Speed:	132 mph (115 kts) (B2M1)
	142 mph (123 kts) (B2M2)
Ceiling:	14,025 ft (4,275 m)
Range:	1,105 miles (960 nm) (B2M1)
	1,094 miles (950 nm) (B2M2)
Engine:	650-hp Mitsubishi-Hispano-Suiza, 12-cylinder water-cooled
Armament:	One fixed forward-firing 7.7mm machine gun and one dorsal flexible 7.7mm machine gun and one Type 91 torpedo or one 1,763 lb (800 kg) bomb

NAKAJIMA NAVY TYPE 3 CARRIER FIGHTER (A1N1)

Wing Span:	31 ft 9 in. (9.68 m)
Length:	21 ft 3½ in. (6.49 m)
Height:	10 ft 8 in. (3.25 m)
Weight Empty:	2,094 lb (950 kg)
Weight Loaded:	3,196 lb (1,450 kg)
Max. Speed:	148 mph (129 kts)
Ceiling:	24,409 ft (7,440 m)
Endurance:	2½ hours
Engine:	730-hp Nakajima Hikari 1, 9-cylinder, air-cooled radial
Armament:	Two forward-firing 7.7 mm machine guns, two 66 lb (30 kg) bombs

NAVY TYPE 13 TRAINER (K1Y1 AND 2)

Wing Span:	33 ft 5¼ in. (10.20 m) (K1Y1&2)
Length:	25 ft 11 in. (7.9 m) (K1Y1)
	28 ft 5¾ in. (8.68 m) (K1Y2)
Height:	10 ft 4 in. (3.15 m) (K1Y1)
	11 ft 4½ in. (3.47 m) (K1Y2)
Weight Empty:	1,477 lb (670 kg) (K1Y1)
	1,922 lb (872 kg) (K1Y2)
Weight Loaded:	2,046 lb (928 kg) (K1Y1)
	2,328 lb (1,056 kg) (K1Y2)
Max. Speed:	89 mph (77 kts) (K1Y1)
	80.6 mph (70 knts) (K1Y2)
Ceiling:	9,843 ft (3,000 m) (K1Y1)
	6,562 ft (2,000 m) (KiY2)

Endurance:	3 hours
Range:	Not known
Engine:	One 130-hp Gausden Benz 6-cylinder, water-cooled, in-line
Crew:	Two – instructor and student
Armament:	None

MITSIBUSHI 1MF1-5 (NAVY TYPE 10 CARRIER FIGHTER)

Wing Span:	30 ft 6 in. (9.3 m)
Length:	22 ft (6.70 m)
Height:	9 ft 8 in. (2.94 m)
Weight Empty:	1,741 lb (790 kg)
Weight Loaded:	2,513 lb (1,140 kg)
Max. Speed:	147 mph (128 kts)
Ceiling:	22,965 ft (7,000 m)
Endurance:	2½ hours
Engine:	300-hp Mitsubishi-Hi, 8-cylinder, water-cooled
Armament:	Two fixed forward-firing 7.7 mm machine guns

MITSUBISHI 2MR1 (NAVY TYPE 10 CARRIER RECONNAISSANCE)

Wing Span:	39 ft 6 in. (12.03 m)
Length:	26 ft (7.90 m)
Height:	9 ft 6 in. (2.89 m)
Weight Empty:	2,160 lb (980 kg)
Weight Loaded:	2,910 lb (1,320 kg)
Max. Speed:	127 mph (110 kts)
Ceiling:	22,965 ft (7,000 m)
Endurance:	3½ hours
Engine:	300-hp Mitsubishi-Hi, 8-cylinder, water-cooled
Armament:	Two fixed forward-firing 7.7 mm machine guns and twin dorsal 7.7 mm machine guns

MITSUBISHI 1MT1N (NAVY TYPE 10 CARRIER TORPEDO)

Wing Span:	43 ft 6 in. (13.26 m)
Length:	32 ft 1 in. (9.79 m)
Height:	14 ft 7½ in. (4.45 m)
Weight Empty:	3,020 lb (1,370 kg)
Weight Loaded:	5,511 lb (2,500 kg)
Max. Speed:	130 mph (113 kts)
Ceiling:	19,685 ft (6,000 m)
Endurance:	3½ hours
Engine:	450-hp Napier Lion, 12-cylinder, water-cooled
Armament:	One torpedo

MITSUBISHI NAVY TYPE 13-1/2 CARRIER ATTACK AIRCRAFT (B1M1)

Wing Span:	8 ft 5½ in. (14.76 m) (13-1&2).
Length:	32 ft 1 in. (9.79 m) (13-1)
	33 ft (10.06 m) (13-2)
Height:	11 ft 6 in. (3.50 m) (13-1)
	11 ft 6½ in. (3.52 m) (13-2)

Weight Empty:	3,179 lb (1,442 kg) (13-1)
	3,891 lb (1,765 kg) (13-2)
Weight Loaded:	5,945 lb (2,697 kg) (13-1)
	6,283 lb (2,850 kg) (13-2)
Max. Speed:	130 mph (113 kts) (13-1)
Ceiling:	14,763 ft (4,500 m) (13-1&2)
Endurance:	2½ hours
Engine:	450-hp Napier Lion, 12-cylinder, water-cooled. 450-hp Mitsubishi Type Hi 12-cylinder, water-cooled
Armament:	Twin dorsal flexible 7.7 mm machine guns (13-1), two fixed forward firing and twin dorsal 7.7 mm machine guns (13-2), one 18-inch torpedo or two 529 lb (240 kg) bombs

Nakajima Army Type 91-1/-2 Fighter

Wing Span:	36 ft 1 in. (11.00 m) (91-1/91-2)
Length:	23 ft 10¼ in. (7.27 m) (91-1)
	23 ft 11¼ in. (7.30 m) (91-2)
Height:	9 ft 2 in. (2.79 m) (91-2)
	9 ft (3 m) (91-2)
Weight Empty:	2,370 lb (1,075 kg) (91-1/91-2)
Weight Loaded:	3,373 lb (1,530 kg) (91-1/91-2)
Max. Speed:	187 mph (162 knts) (91-1/91-2)
Ceiling:	29,527 ft (9,000 m) (91-1/91-2)
Endurance:	2 hours
Range:	Not known
Engine:	450/520-hp Nakajima Jupiter VII, 9-cylinder, air-cooled radial, (91-1)
	460/580-hp Nakajima Kotobuki 2 9-cylinder, air-cooled radial (91-2)
Armament:	Two fixed forward-firing 7.7 mm machine guns

Nakajima Ki-4.

Nakajima Army Type 94 reconnaissance aircraft (Ki-4).

Tachikawa Ki-6.

Yokosho E5K1.

Aichi D3A-35 taking off from an unknown aircraft carrier.

Aichi D3A-19s on an unknown airfield.

A rare airborne shot of a Yokosho E6Y1.

Mitsubishi A5M4-K trainer.

Mitsubishi A5M.

Kawasaki Army Type 93 light bomber.

Kawasaki Ki10 II.

Mitsubishi B5M Type 97 carrier attack bomber (Mabel).

Mitsubishi Ki30-2.

Nakajima J1N-14S

Rare air-to-air shot of Mitsubishi Ki-21 in formation.

Kawanishi H6K5.

Kawanishi H8K2.

Kawanishi H8K2 transport version.

Kawanishi E15K-1

Kawanishi E15K-3 on its beaching trolley.

Mitsubishi A6M2 Zero over China.

Mitsubishi A6M3 Zero being prepared for a mission.

The discovery by the Americans of an almost intact A6M3 Zero on the Aleutian Islands. This aircraft was later restored to flying condition by the US Navy.

Kawanishi H3K.

Mitsubishi G3M Type 96 attack bomber being waved off on a mission.

Mitusbishi G3M Type 96 attack bomber over China.

Mitsubishi G4M Type 1 attack bombers carrying out an extremely low level attack on Guadalcanal.

Nakajima Ki49.

Kayaba Ka-1 autogiro.

CHAPTER THREE

1930s

At the end of 1929 the Nakajima company had acquired the rights to build the Vought O2U Corsair from the American company. They had acquired one of the aircraft for research purposes, and were very impressed with its great potential as both a seaplane and a landplane version. Nakajima's chief engineer Kiyoshi Akegawa made a number of modifications, which included extending the top wing by 12½ in. (500 mm) so as to increase the wing area, extending the interplane struts outward and increasing the size and shape of the rudder, giving it a larger, more rounded appearance. The 450-hp Nakijima-built Jupiter VI engine replaced the 420-hp Pratt and Whitney engine.

The prototype, a single-engined, single-float biplane with wingtip floats, was of a fabric covered, wooden construction except for the forward section of the fuselage, which was metal-covered. The wings, which were rearward folding for ease of stowage, were made of wood, as was the tail section, both being fabric-covered. The aircraft appeared at the end of 1930 and was given the Navy designation of Navy Type 90-2-2 Reconnaissance Seaplane (E4N2). After extensive testing, the aircraft was returned to the Nakajima factory for modifications as the aircraft was considered to have insufficient structural strength. It was almost a year before the Navy would finally accept the aircraft and it was put into production with the 450-hp Nakajima Kotobuki 1 engine, followed a short while later with the 500-hp Kotobuki 2 engine.

With all the necessary modifications and upgrades completed, it was placed aboard cruisers and battleships for further testing. It was received with great enthusiasm after it was discovered that it had the manoeuvrability of a fighter and the structural strength of a dive-bomber. Its other advantage was that it was easily converted into a landplane and given the designation 90-2-3 (E4N3), which highlighted its versatility. Five of the E4N2s were converted for use on aircraft carriers and the aircraft remained in service with the Japanese Navy until 1936, when the Nakajima Type 95 Reconnaissance Seaplane, which saw service during the Pacific War, replaced it.

The Nakajima company built eighty E4N2 seaplanes and five carrier planes between 1931 and 1936, whilst the Kawanishi company built sixty-seven seaplanes between 1932 and 1934.

FURTHER SUBMARINE AIRCRAFT DEVELOPMENTS

Another reconnaissance seaplane appeared in 1934, the Watanabe E9W1 Type 96 Model 11. This was a small, two-seat reconnaissance floatplane constructed like its slightly earlier Western contemporaries, but its Hitachi GK2 Tamput engine gave 340 hp, almost twice that of the Salmson in the French MB-411 for example. A significant step-up in performance was underway. Four prototypes of the E9W1 were produced, followed by thirty-two production aircraft.

This was a single-engined, twin-float sesquiplane of metal and wooden construction with fabric-covered surfaces. The crew were housed in two open cockpits, and the aircraft had been designed to be carried aboard the I-Class submarines. Its main feature was that it could be assembled in just two minutes thirty seconds and dismantled in only one minute thirty seconds on the deck of a submarine.

These aircraft, aboard their parent submarines, were used to seek out Chinese ships attempting to run the blockade during the Sino-Japanese conflict of the late 1930s. By

December 1941, when Japan made her surprise attack on the US bases in Hawaii, the Japanese Imperial Navy had eleven submarines in service capable of carrying aircraft. By the end of World War II Japan would have twenty-seven aircraft-carrying submarines.

It was in 1941 that the large cruiser submarines of the I-9 to I-12 Class came into service. They carried the small Watanabe scouting aircraft E9W1, giving the submarine a long radius of action (16,000 miles at 16 knots). The E9W1 was the first aircraft to be designed and built purely by the Watanabe company, and a total of thirty-three of the aircraft were built between 1934 and 1940.

NEW FIGHTERS ARRIVE

Although the Army appeared to be concentrating on developing its bomber force, it was also looking to upgrade its fighter aircraft. The Kawasaki company were well aware of this and although their KDA-3 parasol-wing fighter had been rejected by the Army, it began a design for a new fighter with an all-metal airframe with a fuselage that had a partially-covered metal front section and a fabric-covered rear section.

Known as the KDA-5, the new prototype fighter was powered by a 500-hp Kawasaki-BMW VI engine, and because of its light construction was subjected to very intense structural strength tests. Having passed all the tests, the first flight took place in July 1930, flown by the company's chief test pilot Kambei Tanaka. It was during one of the tests that the KDA-5 became the fastest Japanese aircraft at the time, attaining a speed of 200 mph (173 kt). This was followed some months later by another record, when the aircraft reached a height of 32,808 ft (10,000 m).

These successes were marred when, on one test flight, the aircraft caught fire and the pilot had to bale out. Despite this setback, the second and third prototypes continued with the test flights and it was during one of the high-speed dive tests that it was discovered that the front interplane strut would bend, causing the aircraft to become unstable. This was remedied, but the testing was halted because of the war in Manchuria and the Army's need for a fighter aircraft.

The Army Type 92 Fighter (KDA-5) aircraft, as it was known, went into production almost immediately and was assigned to units operating in Manchuria and Northern China. It was not much-liked by the pilots, mainly because of its unstable handling during take-offs and landing. It was also a very difficult aircraft to maintain in cold weather, and its operating area of Manchuria and Northern China did not help.

At the same time as the KDA-5 was being produced, Mitsubishi submitted two proposals for a short-range reconnaissance aircraft in response to a requirement submitted to them. One was the 2MR7 biplane, which was rejected; the other was the 2MR8, a high-winged parasol. The aircraft was never meant to replace the Kawasaki Type 88 long-range reconnaissance, but to be used as a lightweight, manoeuvrable, short-range aircraft. The 2MR8 made its first flight on 28 March 1931 powered by a 320-hp Mitsubishi A2 engine, and performed well. The second prototype, which followed a month later, was subjected to rigorous structural tests, which ultimately led to a reduction in the wing area and a shorter fuselage.

The third prototype encompassed all the modifications and was powered by an improved version of the A2 engine: the 345-hp Mitsubishi A2. Despite the modifications, the speed of the 2MR8 was below what was required by the Army, so the fourth prototype was fitted with a 400-hp Mitsubishi Type 92 engine that more than exceeded the requirements. The Army then accepted the aircraft and it was put into production as the Army Type 92 Reconnaissance Aircraft (2MR8).

The aircraft saw service during the Sino-Japanese conflict in Manchuria and Northern China, and was used mainly to give close support to the ground troops. The 2MR8 was the first military aircraft to be powered by an engine that had been designed and built in Japan. Between 1930 and 1934, the Mitsubishi company built 130 of the aircraft.

In 1933 the 750-hp Kawasaki-BMW VII engine appeared and was installed as the Type 92-2. No information is available as to whether or not the aircraft's performance was improved, but a total 385 were built between 1930 and 1933.

In 1932, the JNAF also saw the need to replace one of its carrier attack aircraft, the Mitsubishi Navy Type 89, which was performing very poorly. Instead of putting the project out to tender, the JNAF decided to design and build its own at the Yokosho Arsenal. At the same time, the Mitsubishi and Nakajima companies were being approached to create a design for a carrier attack bomber based on the very successful earlier aircraft: the Mitsubishi Type 13.

Yokosuka's chief designer/engineer, Tamefumi Suzuki, produced a highly modified version of the Type 13, which had a welded tubular steel fuselage and was fitted with a 600-hp Type 91 engine. The prototype was sent to the Navy for test and evaluation and, after extensive testing, was returned to the factory. The results of the tests were very disappointing, highlighting the stability and control of the aircraft as a couple of problems, but the unreliability of the engine was a major factor. Despite these setbacks, when compared with the aircraft from the Mitsubishi and Nakajima factories its problems were insignificant.

Modifications were quickly made and after more testing by the Navy, the Yokosuka Navy Type 92 Carrier Attack Aircraft (B3Y1) as it was designated, was accepted by the Navy and put into production.

The Aichi company was tasked with the production of the aircraft, but as demand grew the Watanabe company and the Hiro Arsenal were also assigned to build the B3Y1. The first of the production aircraft were quickly assigned to JNAF units, and were used very successfully in low-level bombing attacks against small targets during the early stages of the Sino-Japanese conflict. Unfortunately, engine problems dogged the aircraft during its short lifetime, and they were often grounded because of these. It did remain operational, however, during the first few months of the Pacific War, but then was replaced by the Type 96 Carrier Attack Aircraft built by the Aichi and Kusho companies.

A total of 129 Navy Type 92 Carrier Attack Aircraft (B3Y1) were built between 1932 and 1936, one prototype by Yokusho, seventy-five by Aichi, twenty-three by Watanabe, and thirty by the Hiro Arsenal.

A GROWING NEED FOR TRAINERS

As the aircraft started to roll off the production lines in ever-increasing numbers, so there was a need for more pilots. The Navy realised this, and in 1933 approached the Yokosuka Arsenal to design a new intermediate trainer to replace the Type 91 that had been built two years previously. Working with the Kawanishi company, they produced an improved version of the Type 91, the K5Y1.

This was a sesquiplane that had an increased dihedral and sweep of the wings and re-designed upper wing, which was mounted closer to the top of the fuselage. The tail surfaces were also redesigned and were of wooden construction and fabric-covered. The fuselage was constructed of wood and metal, to which was fitted a fixed undercarriage and, later, floats. It was powered by a 340-hp Hitachi Amakaze 11, 9-cylinder air-cooled radial engine, which gave the prototype a top speed of 132 mph (115 knts).

After extensive trials both at the Kawanishi company and the Navy, the Navy Type 93 Intermediate Trainer (K5Y1), as it was designated, went into production in January 1934. There were two versions, the K5Y1, which was a land model, and the K5Y2, which was a seaplane fitted with twin floats.

The Kawanishi company built sixty of both types, but the need for trainers was such that seven other companies, Mitsubishi, Nakajima, Hitachi, Watanabe, Nippon, Kokusho and Fuiji, were tasked with producing the aircraft. It served the Navy throughout the Pacific War, and was given the codename 'Willow' by the Allies.

The K5Y1 was built in larger number than any other Japanese training aircraft and between 1933 and 1945 a total of 5,770 K5Ys were built by various manufacturers.

The Navy then approached both the Aichi and Nakajima companies with regard to the design and development of a reconnaissance seaplane that could be catapult-launched from a battleship or cruiser. The Aichi company sought the help of the Heinkel company and purchased a Heinkel HD 56. With a few minor modifications, the aircraft was submitted to the Navy as the Navy Type 90-1 Reconnaissance Seaplane (E3A1). The Nakajima and Yokosho/Kawanishi companies also submitted aircraft, but it was the Aichi company's aircraft that was selected and put into production.

Powered by a 200-hp Wright Whirlwind engine, the biplane had simple interplane struts that supported the wings without the use of bracing wires and had a very strong and robust fuselage. A couple of things that concerned the Navy were the aircraft's lack of speed and its short range. This was resolved by changing the engine for a 340-hp Type 90 Gasuden Tempu 9-cylinder, air-cooled, radial. The wingspan was also shortened by almost two feet, and the struts that supported the two floats were moved inwards by 12 in. as a consequence.

With the modifications made, the Type 90-1 came off the production line and some of the first were assigned to cruisers who were involved in the Sino-Japanese conflict. Production was halted when just twelve of the aircraft had been built. This was because of the arrival of the Nakajima 90-2-2 Reconnaissance Seaplane, which was infinitely better than the Type 90-1.

The construction of the steel tube fuselage of the Type 90-3 Reconnaissance Seaplane prompted the Hiro Arsenal to take it one step further and produce an all-metal flying boat. The initial design in 1931 began as a twin-engined monoplane flying boat that was to replace the Type 15 biplane and Type 89 flying boats. This design was to produce the first in a long line of prototypes that were to differ in a variety of ways in an effort to secure acceptance by the Navy.

The Hiro Arsenal had built one prototype three-engined flying boat in 1930 called the Type 90-1 that had never gone into production. The new design was a scaled down version of this aircraft but was powered by two engines. Two 500-hp licence-built Pratt and Whitney/Myojo engines powered the Type 91-1, as it was called, but these engines were replaced by 760-hp versions. Over the next two years a variety of changes took place, so many in fact that the design and production of the aircraft were terminated in 1937 because the whole idea had become obsolete.

However there were a number of the aircraft produced during this period and two models were in fact put into short production. The Type 91-1, also known as the H4H1, was a high-winged monoplane fitted with two 600-hp Type 91-2 12-cylinder W-type, water-cooled engines that turned four-bladed wooden propellers. It had a two-step hull and was of an all-metal construction, including the wings and tail section. The second model, the H4H2, was also a high-winged monoplane, but was fitted with two 760-hp Myojo 9-cylinder, air-cooled radial engines. The tail section was different to that of the H4H1 inasmuch as the two rudders were situated slightly higher on the elevator. The H4H2 had the Junkers double wing-type flaps, but neither of the aircraft were considered to be stable while on water, and they did not ride well in heavy seas.

Kawanishi took over the production of the H4H series, as they were known, and produced a total of forty-seven H4H1 and H2s. During the Sino-Japanese conflict both models were used extensively throughout the conflict as transport, mail carriers and cargo: the first Type 91 Flying Boats to be used under wartime conditions.

In the spring of 1934, the design of a new fighter appeared, the Navy Type 95 Carrier Fighter (A4N1). It was powered by the new, powerful Nakajima Hikari 1 engine, which gave it the edge over all the other fighters. The first of the prototypes appeared at the end of 1934, but it wasn't until the January of 1936 that the aircraft was accepted by the Navy. It was immediately put into production as a replacement for the Type 90 Carrier Fighter, which was being phased out.

Because of its increased speed and manoeuvrability, the A4N1 was soon in action in the Sino-Japanese conflict, being used for air defence of Japanese bases. Later they were used as short-range scouting aircraft and, when fitted with 60 kg bombs under the wings, for close air support roles. The A4N1 was to be the last of the biplane fighters, and was later replaced with the Mitsubishi Type 96 Carrier Fighter (A5M). Between 1935 and 1940 a total of 221 Type 95 Carrier Fighters (A4N1) were built.

TOO MANY FOREIGN DESIGNS?

As the Navy and the Army increased their air strengths, they were still dependent on the designs of America and Europe. The number of Japanese-designed aircraft that could be put into production were very few and far between. Then, in 1929, the Army put out for tender the need for a new fighter and invited a number of Japanese companies to enter the competition. The three major companies, Mitsubishi, Nakajima and Kawasaki, put forward designs, and all three were single-seat, high-wing monoplanes but not one was a high-performance aircraft. Three prototypes were built and delivered to the Army for test and evaluation, and all three failed. The Mitsubishi model broke up in the air during a dive test, killing the pilot, while the other two failed to meet the standard required. Nakajima, however, persevered with its Model NC, and in 1930 produced what was to become the Nakajima Type 91 fighter. The aircraft passed all the evaluation tests and went into production as the Army's main fighter. A total of 450 were built.

There were a number of prototype trainers produced over the next few years but most did not get past the design stage, and those that did were not accepted by the Navy for a variety of reasons. One private venture, however, by Mitsubishi, for a single-engine crew trainer, had been produced in 1928. The designer was Herbert Smith and he was asked to design an aircraft with accommodation for a pilot, an instructor and three pupils. Initially it received no interest from either the Army or the Navy, but this design was revived in the 1930s when the need for crew trainers became a priority. The company had built a civilian version and it was this that was handed over to the Army for evaluation. They rejected it for a number of reasons, including its manoeuvrability, a factor which was becoming an obsession with the Army, but the Navy picked it up when a Bristol Jupiter VI radial engine, built by Nakajima, was fitted. Fortunately the Navy were not so obsessed with manoeuvrability in their aircraft as the Army were, and after extensive testing it was accepted and given the designation Type 90 Crew Trainer (K3M1).

Between 1930 and 1941 three versions were produced under the designations K3M1-3. In total 624 of these aircraft were built, seventy-six by Mitsubishi, 247 by Aichi and 301 by Watanabe. During the war the K3M was used extensively for navigation, gunnery, photographic and radio communication training and was still in use when Japan capitulated at the end of the Pacific War.

The Japanese Navy decided in 1932 that a replacement for their Kawanishi Navy Type 90-3 Reconnaissance Seaplane was well overdue. They issued a specification for a three-seat, long-range reconnaissance seaplane and approached two of the major companies, Aichi and Kawanishi, to put forward proposals.

The first prototype to appear was that of the Kawanishi E7K, as it was designated, in February 1933. It was a twin-float biplane powered by a 500-hp Hiro Type 91 12-cylinder liquid-cooled engine. It carried a crew of three – pilot, observer and radio operator/gunner – all situated in individual open cockpits. Armament consisted of two forward-firing 7.7 mm machine guns, one flexible rearward-firing 7.7 mm machine gun in the rear cockpit and one flexible downward-firing 7.7 mm machine gun, also mounted in the rear cockpit. Provision was also made for bomb racks to be mounted under the lower wing centre section, to carry four 66 lb (30 kg) or two 132 lb (60 kg) bombs.

The manufacturers' trials were completed in May 1933 and then handed over to the Navy for test and evaluation. The Aichi company had also completed their trials of Aichi AB-6, and that too was sent to the Navy for test and evaluation. With tests completed the Navy selected the Kawanishi E7K1, as it was superior in almost every way to the Aichi AB-6. The Navy ordered a second prototype and that arrived in December 1933. After further testing, the Navy placed an order for the aircraft, designating it the Navy Type 94 Reconnaissance Seaplane Model 1 (E7K1).

The first models off the production line were fitted with the 500-hp Hiro Type 91 engine, but with the development of an uprated version, the engine was changed to the 600-hp Type 91. The E7K1 was one of the rare aircraft that was loved by all those who flew it, mainly because of its reliability, manoeuvrability and ease of handling. It was assigned to operate from battleships, cruisers, seaplane tenders and land bases. Over the next three years the Kawanishi company built a total of 183 of the aircraft, then production of the E7K1 was handed over to the Nippon company. The Kawanishi company then concentrated on the next version, the E7K2, which was powered by an 870-hp Mitsubishi Zuisei 11, 14-cylinder, air-cooled, radial engine. The first prototype of this model appeared in August 1938 and after test and evaluation went into production in November 1938.

At the beginning of the Pacific War the E7K2, codenamed 'Alf' by the Allies, was being used on anti-submarine patrols, reconnaissance missions, convoy escort duties and training. It remained in front-line service with the JNAF until 1943. Between 1938 and 1941, a total of 344 E7K2s were built by the Kawanishi and Nippon companies, bringing the total built, including the prototypes, to 530.

At the same time as the E7K1 was being designed, another aircraft was appearing on the drawing boards: the Kawanishi H6K. This was in response to a request by the Navy for an experimental large flying boat. The company came up with two proposals, one that had three engines known as Type R and the other Type Q with four engines, both monoplanes. A number of tests were carried out using models in wind tunnels and water tanks, but none came up to the requirements laid down by the Navy.

At the beginning of 1933, Mitsubishi were tasked with designing and building a light bomber and they used the Junkers K37 as a template for their design. The aircraft they produced was known as the Ki-2, the Army Type 93 (Ki-2) Light Bomber. It was powered by two 450-hp Nakajima Jupiter engines and carried a crew of three. Two prototypes were produced, the first appearing in May 1933. Sent to the Army's test and evaluation centre at Kagamigahara, the first prototype was well received, but tragedy struck when, during a second test, the aircraft crashed when it stalled, causing the fuselage to break just aft of the wing roots, killing all the crew.

The fuselage of the second prototype was immediately strengthened and the tapering altered significantly. With these modifications, the aircraft was put through extensive testing and was finally accepted and put into production as the Type 93-1 (Ki-2-1). By the end of 1936, 113 of this type had been built, the majority of which saw action during the Sino-Japanese conflict in Manchuria and northern China.

The Japanese aircraft manufacturers were still heavily reliant on the Western world for their aircraft and were still copying various aircraft designs. But it was a steep learning curve for the young Japanese designers and engineers who were rapidly coming into their own. By the mid-1930s the first of the original designs started to appear. Engines, built under licence from the Western world, were slowly being replaced by Japanese-designed and -built engines. The first of these was the 450-hp Kotobuki, built by the Nakajima company. By the mid-1930s Japanese engines were replacing those from the Western world.

The Army was so delighted with this aircraft that it asked Kawasaki to replace a light bomber that had been built by Mitsubishi. The bomber, the Type 87, was found to be unsuitable for front-line bombing and so Kawasaki was asked to convert a Type 88 reconnaissance aircraft into a Type 88 light bomber. It was not the success that had been hoped for. The Type 88 reconnaissance had a good reputation with pilots because of its reliability and ability to absorb heavy punishment. These characteristics remained with the Type 88 light bomber, but because of its additional bomb load of 440 lbs (204 kg), it became slow and ponderous. Nevertheless, over 400 were built and remained in service until the mid-1930s.

With the ongoing problems with the Navy Type 92 Carrier Attack Bomber (B3Y1) continuing to mount, the Navy urgently needed a reliable replacement. In 1934 they issued a requirement to the various aircraft manufacturers, inviting them to come up with proposals. Mitsubishi, Nakajima and the Dai-Ichi Kaigun Koku Gijittsho (the First Naval Air Technical Arsenal) all submitted proposals, and it was the latter whose design proposal was chosen and a prototype ordered. This was a two-seat equal-span biplane, of an all-metal construction with fabric-covered control surfaces and a fixed undercarriage. A 600-hp Hiro Type 91 liquid-cooled engine powered the first prototype and tests were carried out in October 1935. A 640-hp Nakajima Kotobuki 3 9-cylinder air-cooled radial powered the second and third prototypes, while the fourth and fifth prototypes were powered with the 840-hp Nakajima Hikari 2 9-cylinder, air-cooled radial engines.

The Naval test and evaluation trials were carried out with the aircraft competing against each other, the Mitsubishi Ka-12 and the Nakajima B4N1 being the other competitors. The B4Y1 of Yokosuka was found to be superior in almost every way to the other two and production orders were placed in November 1936. Designated the Navy Type 96 Carrier Attack Bomber (B4Y1), it had an open cockpit for the pilot and an enclosed one for the radio operator/gunner.

The aircraft was operated from aircraft carriers, and saw action in both the Sino-Japanese conflict and at the beginning of the Pacific War. It was soon realised that it was no match for the heavily-armed Allied fighters and was relegated to training duties. The Yokosuka, Nakajima, Mitsubishi and Hiro companies built a total of 205 B4Ys between 1935 and 1938.

The need for a new reconnaissance aircraft to replace the now ageing Mitsubishi Type 92 parasol-monoplane prompted the Army, in 1933, to approach the Nakajima company. The specifications required that the aircraft be of lightweight manufacture, be as manoeuvrable as a fighter and have an all-metal fuselage of monocoque construction. The first of three prototypes appeared in March 1934, followed by the other two in the following months. As this was a contract with a civilian company, the Army maintained a close involvement in every stage of the design and construction.

The first test flights went well, with a number of relatively minor modifications recommended. One of these was to extend the length of the fuselage to improve stability and manoeuvrability. The aircraft was then put into production and given the designation of Nakajima Army Type 94 Reconnaissance Aircraft (Ki-4).

Powered by a 600-hp Nakajima Ha-8 9-cylinder, radial engine, the Type 94 was armed with two fixed forward-firing 7.7 mm machine guns and one flexible dorsal 7.7 mm machine gun. The first aircraft off the production line were fitted with wheel spats, but these were removed on later models. Other prototypes had bomb racks fitted beneath the wings and low-pressure tyres. One of the prototypes was fitted with a centreline float and two wingtip floats, while another had twin floats fitted, but neither went into production. Some of the production models were fitted with flotation bags attached to the sides of the fuselage in case they had to ditch while over water, but the bags did not become part of the standard equipment.

During the Sino-Japanese conflict, the Type 94 was used for general reconnaissance duties, giving close air support to ground troops. On the odd occasion, they were involved in light bombing missions and low-level communications where they dropped message containers and collected information. These types of mission made the aircraft extremely vulnerable to attack from ground fire and a large number of the aircraft were lost carrying out these types of missions.

It was a very popular and forgiving aircraft with aircrew, and ground crew found it very easy to maintain. It also had a wide range of combat capabilities and was seen all over the place during the conflict. It was also the last biplane the Army used for reconnaissance purposes and a total of 1,032 of the aircraft were built, 849 by the Nakajima company, fifty-seven by Tachikawa and 126 by the Manshu company.

A competition was set up to find a design for the Type 2 and Type 15 two-seat reconnaissance aircraft. This time only Aichi and Nakajima entered and they both produced copies of European and American aircraft. Aichi produced a modified version of the Heinkel HD56, a twin-float floatplane powered by a 300-hp Amakaze engine designated the Type 90-I (E3A1). Nakajima produced a modified copy of the Vought Corsair, the Model NJ, and a single-float biplane powered by a 420-hp Bristol Jupiter IV engine and given the designation Type 90-II (E4NS3).

The Yokosuka Arsenal was then approached by the Navy to design a replacement reconnaissance aircraft for its ageing Type 14 seaplane. They, in turn, passed the design to the Kawanishi company to build the aircraft, which produced a twin-float biplane given the designation Type 90-III Reconnaissance (E5K1). Initially the Navy was delighted with the aircraft, but after extensive testing and evaluation its performance fell far short of their expectations and only seventeen were built before production was halted.

With aviation gathering momentum, the need to train pilots became more and more demanding, so much so that the Army approached a small private company, Tachikawa, to design and build them a trainer. This had come about after the Army had tested the company's R-5 primary trainer, and although it was too small for their requirements, the tests revealed a sturdily-built aircraft that was simple to fly. The specifications laid down by the Army were that the biplane trainer had to be able to accommodate different engines, which enabled it to be used as a primary trainer and an intermediate trainer. The aircraft had to be fitted with blind-flying instrumentation, have a top speed of 137 mph (220 km/h), an endurance of three and a half hours and be capable of carrying out 12 g manoeuvres.

By the end of 1934 three prototypes of the Tachikawa Ki-9 had been built, and the first flight was carried out in January 1934. The first test flight threw up problems with the aircraft's manoeuvrability because of the heaviness of the controls. It was discovered that the centre of gravity was far too forward and after some modifications the problem was resolved. The main undercarriage shock absorbers were found to be much too hard and were replaced with softer ones. With the modification completed, the Ki-9 trainer was passed to the Army for test and evaluation.

With the trials completed, the aircraft was put into production, and between 1934 and 1945 2,618 of the aircraft were built. Just towards the end of the war a lighter version was built with a stronger undercarriage and a shorter fuselage. The trainer saw service in the Sino-Japanese conflict and the Pacific War and was codenamed 'Spruce' by the Allies. Even the Manchurian and Thai Air Forces operated some of the aircraft during this period. This small trainer was one of the most resilient aircraft built by the Japanese.

In the early years of the development of the Ki-9, because of problems encountered in the designing, the Army became frustrated with the delay and instructed the Tachikawa company to design another primary trainer; this one was to be stressed for only 6 g and have a loaded weight of less than 2,205 lb (1,000 kg).

Like the Ki-9, the Tachikawa Army Type 95-3 Primary Trainer (Ki-17), as it was designated, had ailerons fitted to both the upper and lower wings, but during flight trials it was discovered that they caused the controls to become over-sensitive and so the upper wing ailerons were removed. Within weeks of the tests, that aircraft was put into production, and between 1935 and 1944 560 of the aircraft were built. This aircraft, like the Ki-9, saw service in the Sino-Japanese conflict and the Pacific War, and was given the codename 'Cedar' by the Allies.

In 1935, the Nakajima company were also involved in producing a trainer at the same time as the Tachikawa company. Given the designation Nakajima Army Type 95-2 Trainer (Ki-6), this was just a modified version of Nakajima-Fokker Super Universal Transport that had been built in 1931 and whose design was then considered to be obsolete. The aircraft had a good reputation for its reliability, ease of maintenance and simplicity to fly. There had been ongoing problems with the wooden wings warping in the high humidity of Japan and the unreliability of the Nakajima-built Jupiter VI engines, but despite this, forty-seven of the aircraft had been built.

Because this was an existing aircraft, no prototype was produced and the production models, known as Ki-6, had oversized low-pressure tyres and were powered by the Nakajima Jupiter VII engine that turned a three-bladed propeller instead of the two-bladed version on the earlier model. The fuselage was constructed of welded steel tube covered in fabric, as was the tail section.

As well as two crewmembers, four crew-training positions were provided for: navigation, radio communication, aerial photography and gunnery. The latter was provided in an open cockpit on top of the fuselage, just behind the trailing edge of the wing, in which a flexible 7.7 mm machine gun was mounted.

Between 1934 and 1936 the Nakajima company built thirty-nine of the aircraft and as well as its training duties, it was sometimes used for transport and liaison missions.

As far back as 1933 it became apparent that there was a need for an aircraft carrier bomber. Two carrier dive-bombers built by the Nakajima company had been produced at the beginning of the 1930s, and neither had stood up to the rigours of operating from the deck of an aircraft carrier. The JNAF put forward specification for a carrier bomber to three companies, Aichi, Nakajima and Yokosuka. Still influenced by Western aircraft designers, Aichi imported a Heinkel He 66 single-seat biplane dive-bomber powered by a 715-hp

Siemens SAM-22B 9-cylinder, air-cooled radial engine. This aircraft was never designed to operate from an aircraft carrier, so using the basic design of the Heinkel they re-designed a stronger undercarriage that would withstand the pounding it would suffer on deck landings, and replaced the engine with a 560-hp Nakajima Kotobuki 2 Kai 1, 9-cylinder, air-cooled radial. They also installed an observer/gunner's seat behind the pilot.

During the trials by the Navy the Aichi D1A, as it was now called, met all the requirements and proved to be more reliable and manoeuvrable than its competitors. A contract was awarded to the Aichi company and the Navy Type 94 Carrier Bomber (Aichi D1A) started to come off the production line. A number of modifications were made to the production model as opposed to the prototype. Both the wings and rudder were given a five-degree sweepback, a Townsend ring was fitted around the engine cylinders and the tailskid was replaced by a fixed tailwheel. Two forward-firing 7.7 mm machine guns and one flexible 7.7 mm rearward-firing machine gun made up the defensive armament. The Aichi D1A could also carry two 66 lb (30 kg) bombs under the wings, and one 551 lb (250 kg) bomb that was attached to a swing-down mechanism mounted beneath the fuselage.

The first 118 D1As that came off the production line were fitted with the 580-hp Nakajima Kotobuki 2 Kai 1 engine, whereas the last forty-four, D1A2s, were fitted with the improved Kotobuki 3 engine. The aircraft saw action during the Sino-Japanese conflict and performed well, but by the time the Pacific War had started the remainder of the 590 that had been built had been relegated to training duties.

On 7 July 1937, the Sino-Japanese conflict flared up again on the mainland of Asia. By this time the Japanese Navy had increased their carrier striking force considerably:

First Carrier Division
Rear-Admiral Shiro Takasu

Carrier *Ryujo* under the command of Captain Katsuo Abe
12 Nakajima Type 95 fighters
15 Nakajima Type 95 dive bombers

Carrier *Hosho* under the command of Captain Ryunosuka Kusaka
9 Nakajima Type 95 fighters
6 Aichi Type 92 attack bombers

Second Carrier Division
Rear-Admiral Rokuro Horie

Carrier *Kaga* under the command of Captain Ayao Inagaki
12 Nakajima Type 90 fighters
12 Nakajima Type 94 dive bombers
12 Kawanishi Type 89 attack bombers
12 Nakajima Type 96 torpedo-bombers

Land Based Naval Air Force
First Combined Air Flotilla based in Shanghai
Captain Mitchitaro Tozuka

Kisarazu Air Corps under the command Captain Ryuzo Takenaka
6 Mitsubishi Type 95 attack bombers
24 Mitsubishi Type 96 attack bombers

Kanoya Air Corps under the command of Captain Sizue Ishii
9 Nakajima Type 95 fighters
18 Mistubishi Type 96 attack bombers

Second Combined Air Flotilla
Rear-Admiral Teizo Mitsunami

Twelfth Air Corps under the command of Captain Osamu Imamura
12 Nakajima Type 95 fighters
12 Nakajima Type 94 dive-bombers
12 Aichi Type 92 attack bombers

Thirteenth Air Corps under the command of Captain Sadatoshi Senda.
12 Mitsubishi Type 96 fighters
12 Mitsubishi Type 96 dive-bombers
12 Mitsubishi Type 96 torpedo-bombers

This gave the Navy a total strength of:

66 carrier-based fighters
51 carrier-based dive-bombers
54 carrier-based attack and torpedo-bombers
48 land-based attack bombers

i.e. a total of 219 combat-ready aircraft

These figures do not include the reconnaissance aircraft and seaplanes that were assigned to either land bases or ships.

A FRESH INCIDENT IN CHINA

On 14 August 1937, the first major incident of the conflict involving the Naval aircraft came about when word came down that Japanese marines in Shanghai were about to be overwhelmed by Chinese troops. Chinese aircraft opened the attack on the garrison there and because the closest airfield to Shanghai was in Chinese hands, it was impossible to get aircraft close to the fighting. The Japanese Navy bombers attacked the Chinese troops after flying from bases on Formosa and Kyushu Island, a distance of 1,250 miles for each attack. These attacks constituted the first transoceanic bombing raids, and stunned the Chinese, who thought that the Japanese had no aircraft in the area. They were taken completely by surprise and the land-based aircraft were joined the following day by aircraft from aircraft carriers. With the advantage gained, the Chinese were forced to withdraw after suffering major losses. The Navy, however, did suffer a number of losses from enemy aircraft and groundfire, but they had made their mark. The Navy also realised the bombers were no match for fighter aircraft, and if long-range bombing missions were to be undertaken, then ideally the bombers required a fighter escort.

On 17 August, a raid on the Chinese mainland included aircraft from the aircraft carrier *Kaga*, and of the twelve Mitsubishi Type 89 attack bombers, only one returned. Flown by Sub-Lieutenant Tanaka, the bullet-ridden, crippled bomber returned to the carrier. Tanaka explained what had happened: the mission had been disastrous because of bad weather, the expected fighter escort never appeared and so the bombers went on alone and were attacked by enemy fighters. He was the only one to survive. Not only had the Navy lost eleven of its bombers, but they had also lost eleven very experienced crewmembers. This incident strengthened the argument for the need for escort fighters.

Over the preceding years, the Japanese Navy and Army had concealed their aircraft and armaments from the world; only the obsolete aircraft warships and heavy guns were allowed to be seen. This lulled the surrounding countries into thinking that Japan was incapable of being able to defend itself, let alone attack another country. By purchasing foreign aircraft and guns, the Japanese military machine was able to gauge the military capability of other countries, and by keeping their rapidly developing armament away from prying eyes, allowed other countries to seriously underestimate Japan's combat strength.

The conflict in China was dragging on and although attempts were made through diplomatic channels to bring the conflict to an end, the Japanese Army hierarchy and some of the warlords in China hampered it. The Japanese Army's influence was such that they forced Prime Minister Konoye to announce that the Japanese government 'will not negotiate with the Nationalist Government of Chiang Kai-shek'. The war continued, and the influence the Army and the Navy had over the government became more and more obvious and the development of their respective air arms continued.

In 1934 the Navy had issued a specification to Aichi, Kawanishi and Mitsubishi, which called for a short-range observation aircraft capable of being catapult-launched from cruisers or battleships and intended to replace the Nakajima E8N1 currently being used. The Mitsubishi company wasted no time in getting to work designing its prototype, which appeared at the beginning of 1936 and was first flown in June the same year. This was a two-seat, single-engine biplane of an all-metal construction with fabric-covered control surfaces. The Mitsubishi F1M1, as it was designated, had wings that were of equal span but with only single interplane struts. It had a central float and two outboard stabilisers and was powered by an 820-hp Nakajima Hikari 1 9-cylinder radial engine.

The first flight tests highlighted a problem when the F1M1 was taxiing or taking off from water, inasmuch as it had a tendency to 'porpoise'. Also, the flight directional stability was very poor. Four more prototypes were produced before both of these problems had been eradicated. The final prototype, the F1M2, was powered by an 875-hp Mitsubishi Zuisei 13, 14-cylinder radial engine. It also had improved visibility for the pilot and redesigned wings that had straight leading edges in place of the elliptical ones that had been a feature of the F1M1. The surface areas of the vertical fin and rudder were significantly increased, which, together with the other modifications, improved the handling characteristics considerably. Armament consisted of two fixed forward-firing 7.7 mm machine guns and one reward-firing flexible 7.7 mm machine gun.

With the test and evaluations by the Navy carried out satisfactorily, the aircraft was put into production as the Mitsubishi Navy Type O Observation Seaplane Model 11 (F1M2). It was soon discovered that the manoeuvrability of the aircraft was far superior to some of the fighter aircraft and it was soon being used as an interceptor, a fighter, a dive-bomber, and for coastal-patrol duties. On the odd occasion it was also employed as an observation aircraft, its original intended role.

Production of the aircraft initially was the responsibility of Mitsubishi, but as the company became busier with the production of other aircraft, the Kokusho Naval Arsenal took over, producing 590 F1M2 aircraft.

The F1M2 was one of the Japanese Navy's most successful multi-role aircraft, although it was never intended as such, and a total of 1,118 were produced. It saw service throughout the Pacific War aboard battleships, cruisers, seaplane tenders and land bases, and was known to the Allies as 'Pete'.

The JNAF had, for some time, been looking for a replacement for the Kawanishi E7K2 three-seat, twin-float seaplane and so issued a specification to three of the main manufacturers, Aichi, Kawanishi and Nakajima. This time the specification called for a two-seat seaplane that had a longer range and higher speed. Soon after issuing the specification, the Navy issued a second specification to the three companies, calling for a three-seat version.

It was the Aichi company that had come up with proposals for the two aircraft, while the Nakajima company concentrated on just the one, the two-seater. Kawanishi concentrated on the three-seat version. All three manufacturers produced prototypes but it was the Aichi E13A1, a three-seat version powered by a 1,060-hp Mitsubishi Kinsei 43 engine, that was selected by the Navy in December 1940. After extensive testing production was started and the Aichi company had produced a total of 133 of the aircraft when it was told to hand over the main production to the Watanabe company. Aichi was then told to concentrate on producing the D3A and D4Y carrier bombers that were urgently needed. The Kokusho Arsenal undertook to build the E13A1, and between 1940 and 1942 built forty-eight of the aircraft. The Watanabe company built the lion's share by building 1,237 E13A1s between 1942 and 1945. In total 1,418 of the aircraft were built.

The E13A1 operated from cruisers and seaplane tenders and was used primarily for reconnaissance missions and anti-shipping patrols. They first saw action when they were used for attacking the Canton-Hankow railway during the Sino-Japanese conflict that was still continuing when the Japanese attacked Pearl Harbor in December 1941.

During the attack on Pearl Harbor, E13A1s, known as 'Jake' to the Allies, flew photo-reconnaissance flights after the raid so that the damage to the American fleet could be assessed. The E13A1 was also used for air-sea rescue work, shipping attacks and – at the very end of the war – Kamikaze (Divine Wind) attacks.

THE 'VAL' DIVE-BOMBER

While the D1A was being built, the Navy had issued a specification calling for a carrier-based monoplane dive-bomber to be built. In the summer of 1936, the three major aircraft manufacturing companies, Aichi, Mitsubishi and Nakajima, were approached and asked to build two prototypes to the Navy's specifications. Aichi looked at the best features of a number of successful Western aircraft designs and selected the low-mounted elliptical wings of the Heinkel He 70. Aichi had considered incorporating a retractable undercarriage, but decided against it on the grounds that the additional weight and complexity of the mechanism did not justify the slightly increased speed they would gain. The D1A was one of the last Japanese naval aircraft to have a fixed undercarriage with spatted wheels.

In December 1937 the first of the two Aichi prototypes appeared, powered by a 710-hp Nakajima Hikari 1 9-cylinder, air-cooled radial engine. Flight trials commenced in January 1938 and were extremely disappointing, encompassing a number of deficiencies. The design of the dive brakes had been taken from those of the Junkers Ju 87, and it was found that they vibrated violently when they were turned through 90 degrees to present a flat surface to the airflow during high-speed dives. The area had to be increased and the mounting strengthened considerably. It was also underpowered, and during wide turns suffered from directional instability and had a tendency, during tight turns, to go into a snap roll. There were, however, plus sides: the aircraft possessed a very strong airframe, was extremely manoeuvrable, and handled well.

The second prototype was produced with all the deficiencies addressed, and the engine was replaced with an 840-hp Mitsubishi Kinsei 3 14-cylinder, air-cooled radial engine covered with a re-designed cowling. The dive brakes had been strengthened considerably, the vertical tail surface enlarged, and the wingspan increased from 46 ft 3¼ in. (14.1 m) to 47 ft 6¾ in. (14.5 m); this, of course, increased the wing area by 21.8 sq. ft (2 sq. m). In an effort to stop the aircraft going into snap rolls during tight turns, the outer sections of the leading wing edges were cambered down. Fitting a large dorsal fin to the vertical tail solved the problem of the directional instability.

In December 1939, Aichi were awarded a contract to build the dive-bomber after extensive testing by Navy test pilots and carrier qualification trials aboard the aircraft carriers *Kaga* and *Akagi*.

Designated the Aichi Navy Type 99 Carrier Bomber Model 11 (D3A1), the first aircraft off the production line had been modified even more, having slightly smaller wings and a more powerful engine, a 1,000-hp Mitsubishi Kinsei 43. Armament consisted of two fixed 7.7 mm forward-firing machine guns and one flexible rearward-firing 7.7 mm machine gun. A 551 lb (250 kg) bomb was slung beneath the fuselage and held by means of two arms that swung downwards and forwards before releasing. Two more 132 lb (60 kg) bombs were mounted beneath the wings in wing racks and outboard of the dive brakes.

The aircraft first saw action in the on-going Sino-Japanese conflict and operated from land bases. On 7 December 1941 the Japanese attacked Pearl Harbor, and the first aircraft to drop a bomb on the Americans was the D3A1. During the following year the D3A1 or 'Val', as it was codenamed by the Allies, was responsible for sinking more Allied fighting ships that any other single type of Axis aircraft. During the battle in the Indian Ocean, the aircraft was largely responsible for the attacks on the British cruisers HMS *Dorsetshire* and *Cornwall* and the aircraft carrier HMS *Hermes*.

As the war progressed and the Allied aircraft became faster and more powerful, the D3A1 started to suffer heavy casualties. During the Battle of the Coral Sea their numbers were decimated, and the loss of the elite crews with them forced the Navy to assign them to land bases in an attempt to reduce the losses.

An improved version, the D3A2 (Model 12), powered by a 1,300-hp Mitsubishi Kinsei 54 engine, made its appearance. This model had larger fuel tanks, giving it a total capacity of 237 gallons (1,079 litres), which increased its operating range considerably. Gradually they replaced the D3A1, but within months they too had been replaced by the Yokosuka D4Y Suisei and relegated to land-based units. When in 1944 the Americans fought to take back the Philippines, the Aichi D3A was in the thick of the fighting, only this time in the role of the kamikaze, and suffered horrendous losses.

At the beginning of 1937, the JNAF had issued a requirement for a primary seaplane trainer as a replacement for their Navy Type 90 Primary Seaplane Trainer that was rapidly becoming obsolete. Three manufacturers had been approached: Kawanishi, Nippi and Watanabe, and it quickly became obvious that there was almost no latitude with regard to the design concept.

The Kawanishi prototype appeared in May 1937 and was given the designation K8K1, followed by the Nippi prototype K8Ni1 and the Watanabe prototype K8W1. All three aircraft were almost identical to look at, but it wasn't until 6 July 1938 that the extensive testing and evaluation by the Navy was completed. The Kawanishi model was selected and production started in June 1940. Given the official designation Navy Type O Primary Trainer (K8K1), the aircraft had a fabric-covered, welded steel tube fuselage, fabric-covered wooden wings and twin metal floats and was powered by a 130-160-hp Gasuden Jimpu 2, 7-cylinder, air-cooled, radial engine. Although it was far superior to the Type 90 trainer it was replacing, only fifteen were ever made because the appearance of the Type 93 Intermediate Trainer made it obsolete.

SUBMARINE-LAUNCHED AIRCRAFT

The Navy had also been starting to show an interest in the idea of placing an aircraft aboard a submarine. This had come about when, back in 1919, they had purchased seven War Prize U-boats from the German Navy, with the intention of adopting the best features into the designs of their own submarines.

A trial had taken place on 6 January 1915 when a Friedrichshafen FF.29 seaplane, with a fifty-seven foot wingspan, was lashed down across the foredeck of the German submarine U-12. Some thirty miles offshore, the U-boat's commander had flooded the forward tanks and released the aircraft, which was able to take off successfully. The aircraft, with von Arnauld de la Periere and his observer Herman Mall aboard, flew along the coast of Kent undetected before returning to Zeebrugge direct, rather than making the agreed rendezvous with the U-12. Despite the success of a number of subsequent trials, the idea was shelved.

With this concept in mind, the Yokosuka arsenal had produced a small, single-seat reconnaissance seaplane based on the design of the British Parnall Peto. Designed by Lt-Com. Jiro Saha and Tamefumi Suzuki, the small biplane, with a wingspan of only twenty-six feet and powered by a 130-hp Kamikaze II engine, had been designed specifically for use aboard a submarine. Built by the Kawanishi company, the Yokosho E6Y1 Type 91 Reconnaissance Seaplane, as it was called, was powered by the 130hp Armstrong Siddeley Mongoose engine. Its predecessor, the Peto, had been successfully tested aboard the ill-fated British M.2 submarine. The first successful testing of the Yokosho E6Y1 was in May

1928, from the submarine I-51. Subsequent tests in 1930 and 1931 were successful and the aircraft was officially adopted as a reconnaissance seaplane for the submarine fleet. Nine of the aircraft were ordered and delivered to the Navy.

The Navy were beginning to increase and improve their Air Arm considerably and ordered an improved Type 3 fighter that had been built by Nakajima. This was the Navy's principal fighter, and information coming back from the Western world told them of the development of much faster and more heavily armed fighter aircraft. Nakajima's chief designer, Takao Yoshido, produced a new model based on the old design of the Type 3, but incorporating many of the features of the American Boeing 69B Fighter that the Navy had imported in 1928. Two prototypes were built and submitted to the Navy for evaluation, but despite the modifications, the Navy rejected it on the grounds that there was insufficient improvement over the Type 3.

In a desperate effort to remain in the running for the new fighter contract, Nakajima used an imported British Bristol Bulldog that had originally been acquired with the hope of using some of the design techniques for a fighter for the Army. The project was given to the designer, Jingo Kurihara, who set about redesigning the Type 3 and incorporating some of the best features of the Bulldog into the new aircraft. Powered by a 580-hp Nakajima Kotobuki 2 engine, it was sent to the Navy for test and evaluation. The Navy saw a tremendous improvement over the Type 3 and placed an immediate order.

The new aircraft was given the designation Type 90 Carrier-based Fighter (A2N1-2) and over the next few years a number of different models appeared, but these differences were only minor. By the beginning of 1936 over 100 had been built and delivered to the Navy. A two-seat trainer was also built, the A3N1, but only in small numbers.

THE MANCHURIAN INCIDENT

The training of young Japanese pilots became a priority when the Manchurian and Shanghai Incidents erupted. The trigger for this war was the blowing up of the Japanese railway line in Manchuria close to the city of Shenyang on 18 September 1931. The Japanese blamed the Chinese for the outrage, but in fact it was the Japanese Army that had blown up the line as a pretext for the invasion of Manchuria. With only a little over 10,000 troops stationed in Manchuria, the Japanese Army dispatched three more squadrons from the Fourth and Seventh Air Wings to Shenyang, to bolster the Sixth Air Wing's three squadrons already stationed there. The Japanese-built aircraft consisted largely of Army Type 87 Heavy Bombers, Type 87 Light Bombers, Type 88 Reconnaissance, and Type 91 Fighters. The remainder were Nieuport fighters, Salmson 2A2 reconnaissance, Junkers K-37 bombers and the Dornier air ambulance.

The Chinese had a small number of French Potez 25 reconnaissance aircraft, but these were captured within days of the war starting and were used by the Japanese Army. This meant that the Chinese Army were unable to mount a single air operation against the Japanese, leaving the control of the sky to the invading army.

With six squadrons operational in Manchuria, a Hiko-tai (Kanto Air Command) was formed to oversee all operations. Three more squadrons were sent to Manchuria to bolster the existing six squadrons. It was decided that now there were nine squadrons, they would establish them as three Air Battalions, consisting of three squadrons each. They were the Tenth Air Battalion (Reconnaissance Squadrons), Eleventh Air Battalion (Fighter Squadrons) and Twelfth Air Battalion (Bomber Squadrons).

The Air Battalions had been formed primarily as support for the ground forces and all the missions were carried out with that in mind. As the war progressed and the Japanese Army marched through Manchuria and into China, the Russians, always fearful of any threats to their borders, moved their ANT TB-3 long-range bombers to their air bases in Primornuli State in the Far East.

The Japanese Army recognised the need for a long-range bomber, but seriously underestimated the specifications that they had laid down when they had put out requests for such an aircraft. Throughout the war in Manchuria, the Army never had a long-range bomber. Fortunately, the conflict in Shanghai was a naval affair and when fighting broke out between Japanese Marines and the local Chinese, the Navy's Air Arm was able to support the ground forces and help bring the incident to a rapid conclusion in just thirty-four days. The seaplane carrier *Notoro* had been lying off Shanghai at the time and immediately hostilities started, they launched their Type 14 and Type 90-3 reconnaissance/ bombers to make sorties over the city. At the end of January the aircraft carriers *Kaga* and *Hosho* arrived, carrying aircraft of the First Air Wing consisting of Type 3 Fighters and Type 13 Torpedo-bombers, a total of seventy-six aircraft. On 1 February attacks from these aircraft commenced and almost daily sorties were carried out on the city and the surrounding area.

At the end of February the Chinese Air Force bases were attacked, and for the first time they were involved in air combat. Nine Chinese aircraft were destroyed on the ground but no Japanese aircraft were shot down, although they did suffer some minor damage.

With the war going well for the Japanese, the Army decided to send five squadrons to support the naval aircraft in their ongoing assault against Shanghai and the surrounding towns. At the beginning of May, opposition against the Japanese Army and Navy had all but disappeared, and on 8 May an armistice agreement was reached. The war had shown to the Japanese the effectiveness of aircraft and established both the Army and Navy Air Wings as a necessary part of the military machine.

With their positions now well established, both the Army and the Navy air sections used their success to 'flex their muscles', and put forward plans to increase the number of aircraft and subsequently the number of squadrons. They also took the opportunity to make the Air Wings independent, becoming a separate part of the military with their own command structures. The Japanese Air Command was structured thus:

The First Air Group

First Air Wing	4 Fighter Squadrons	Kasimigaura
Second Air Wing	2 Reconnaissance Squadrons	Kasimigaura
Third Air Wing	3 Reconnaissance Squadrons	Yokaichi
Seventh Air Wing	2 Light Bomber Squadrons	Hamamatsu
Seventh Air Wing	2 Heavy Bomber Squadrons	Hamamatsu
Thirteenth Air Wing	3 Fighter Squadrons	Kakogawa

The Second Air Group

Sixth Air Wing	1 Fighter Squadron	Pyongyang
Sixth Air Wing	2 Light Bomber Squadrons	Pyongyang
Ninth Air Wing	2 Fighter Squadrons	Hoelyong
Ninth Air Wing	2 Light Bomber Squadrons	Hoelyong

The Third Air Group

Eighth Air Wing	1 Fighter Squadron	Pingtung
Eighth Air Wing	1 Light Bomber Squadron	Pingtung
Fourteenth Air Wing	2 Heavy Bomber Squadrons	Chaii
Fourth Air Wing	2 Fighter Squadrons	Tachiarai
Fourth Air Wing	2 Reconnaissance Squadrons	Tachiarai
Fifth Air Wing	2 Fighter Squadrons	Tachikawa
Fifth Air Wing	2 Reconnaissance Squadrons	Tachikawa

The Kanto Command
The Joint Air Group

Tenth Air Wing	3 Light Bomber Squadrons	Tsitsihar
Eleventh Air Wing	4 Fighter Squadrons	Kharbin
Twelfth Air Wing	4 Heavy Bomber Squadrons	Kungchuling
Fifteenth Air Wing	3 Reconnaissance Squadrons	Changchun
Sixteenth Air Wing	2 Fighter Squadrons	Mutanchiang
Sixteenth Air Wing	2 Light Bomber Squadrons	Mutanchiang

The Army was slowly replacing all its foreign aircraft with Japanese-designed ones and was increasing the number of bomber squadrons.

In 1932, the JNAF had begun a re-equipment programme and included was the need to rid itself of the reliance upon Western aircraft manufacturers for aircraft. They put out to tender a requirement for a single-seat fighter that could be used both on land and aboard aircraft carriers. Two designs were put forward based on a seven-point requirement, one from Mitsubishi (a low-wing monoplane) and one from Nakajima (a parasol monoplane). Both the Mitsubishi design and the Nakajima design were rejected, which caused the Navy to order another biplane, the Nakajima A4N1, but only as a temporary measure. The Nakajima Navy Type 95 Carrier Fighter (A4N1) wasn't put into production until January 1936, and was powered by a 730-hp Nakajima Hikari 1, 9-cylinder, air-cooled, radial engine, giving the aircraft a top speed of 218 mph (190 knots). It was of an all-metal construction but with fabric-covered control surfaces. Its armament consisted of two fixed, forward-firing 7.7 mm machine guns, and provision was made to carry two 66 lb or one 132 lb (30 kg or 60 kg) bombs beneath the wings. There were a number of problems with the prototype, and a number of modifications had to be made before the Navy would accept it.

When the Type 95 did enter service, it arrived at the time of the Sino-Japanese conflict and was used for air defence over Japanese bases. The A4N1 was to be the last of the biplane fighter aircraft used by the Navy and even as it was involved in the conflict, a new and revolutionary fighter was already under development at the Mitsubishi plant. The Type 96 fighter, the A5M, known as 'Claude' to the Allies, replaced the A4N1 less than a year later. Over 200 were built between 1935 and 1940.

In the autumn of 1933 the Japanese Navy put out a tender for a two-seat reconnaissance seaplane to replace their Type 90-2 (E4N2). It had to be capable of operating from aircraft tenders, battleships and cruisers. Nakajima had built the E4N2, and decided to enter the competition by submitting an updated version of the aircraft. Two other manufacturers, Aichi and Kawanishi, had entered the competition, both looking toward designing low-wing monoplanes.

The Nakajima E8N1, as it was known, differed from the E4N2 in having a smaller wing area, an increased sweep on the upper wing, and a taller vertical tail surface without the dorsal fin. It was powered by a 580-hp Nakajima Kotobuki 2 KAI I 9-cylinder radial engine. The first of seven prototypes was produced in March 1934, and over the next year was tested against the two other prototypes produced by the Aichi and Kawanishi companies.

The E8N1 proved to be the better all-round, being superior in all the handling and manoeuvring tests it was subjected to. It was put into production and given the designation Navy Type 95 Reconnaissance Seaplane Model 1.

It was operated from battleships and cruisers, and during the Sino-Japanese War it saw action against Chinese fighter aircraft and distinguished itself by shooting down a number of them during attacks on Shanghai.

The Nakajima factory produced 707 E8N1s, while the Kawanishi factory produced forty-eight between 1934 and 1940.

The replacement, the Mitsubishi Type 96 Ka-14 (A5M) prototype, was designed by Jiro Horikoshi, who produced an inverted gull wing monoplane with a fixed spatted undercarriage. In an effort to minimise drag, the airframe consisted of small cross-sections with flush-riveted, stressed aluminium covering it. The prototype was powered by a 550-hp Nakajima Kotobuki 5 radial engine, and during initial tests far exceeded all expectations. There were problems, however, during flight, as the aircraft suffered from pitch oscillations and a tendency to balloon when landing. The problems resulted in the second prototype having considerable modifications, such as the dihedral outboard to the wingtips being reduced by seven degrees, with split trailing-edge flaps being installed on the centre section of the wing. The engine was replaced with a direct-drive Nakajima Kotokubi 3 radial.

The arrival of the Mitsubishi A5M1, as it became known, suited requirements perfectly. The outbreak of the Second Sino-Japanese War saw the arrival of the A5M2a. This aircraft was powered by a Nakajima Kotobuki 2 KAI 3A engine, and almost immediately gave the Japanese air superiority. Improvements to the A5M2a resulted in the production of the A5M2b, which had an enclosed cockpit, a 640-hp Nakajima Kotokubi 3 engine and a three-bladed propeller. The A5M2b was superior to anything the Chinese had, and as the war progressed so did the development of the A5M2.

One of the developments was the production of a prototype A5M3, which was powered by a 610-hp Hispano Suiza engine that had been brought over from France. The policy of the Japanese Navy was to not rely on Western aircraft and engines, and so they rejected the aircraft, but with the production of the A5M4, powered by a 710-hp Nakajima Kotobuki 41 radial engine and with a thirty-five gallon (160-litre) ventral drop tank, the Navy's attitude changed. Production orders for the Model 24, as it became known, were placed, and in 1938 as the first of the aircraft came from the manufacturers, they were sent to combat units in China, where they continued to decimate the Chinese Air Force. Of all the A5M series of aircraft, the A5M4 was built in larger numbers than any of the others. Initially the A5M4 made up the bulk of Japanese carrier fighters, but by the onset of the Pacific War another Japanese carrier fighter, the legendary Mitsubishi A6M 'Zeke', had superseded it.

Between 1935 and 1940, the Mitsubishi company produced 791 of the aircraft; the Watanabe company then produced a further thirty-nine A5M4 and the Kokusho company another 264, making a total of 1,094 A5Ms built.

With the relative success of the Type 92 Fighter still in mind, the Army awarded the Kawasaki company another contract, this time to design and build a single-engined light bomber. Work started in September 1932, and the Ki-3 prototype appeared seven months later in April 1933, powered by an 800-hp Kawasaki-BMW IX engine. Using the same light alloy construction, the front section of the fuselage was of a metal covering, while the remaining section was fabric-covered. The crew of two sat in open cockpits, and the Ki-3 was armed with one fuselage-mounted, forward-firing 7.7 mm machine gun, and twin flexible, dorsal-mounted 7.7 mm machine guns. It carried a 1,102 lb (500 kg) bomb load.

Two more prototypes were built before the Army accepted them, and the aircraft was put into production with the designation Army Type 93 Single-Engined Light Bomber. Once the aircraft started to roll off the production line, they were assigned to units in Northern China and Manchuria, where they carried out bombing, reconnaissance and cargo missions in support of the ground troops. Because of the demand for the aircraft, the Tachikawa company undertook the building of forty. A total of 243 were built between 1933 and 1935, when production was halted due to continuing engine problems.

Kawasaki produced the last of the Japanese Army's biplane fighters in 1935, when they were asked by the Koku Hombu to design and build a fighter aircraft that was superior to those of the Western world. Kawasaki's chief designer Takeo Doi designed a biplane of unequal span, which had ailerons fitted only to the upper wing and was powered by a Kawasaki Ha-9-IIa engine. The Kawasaki Ki-10 prototype, as it was known, was completed in March 1935, followed one month later by a second.

Flight tests took place in April and surpassed all that was asked of them. The Ki-10 was superior in both speed and manoeuvrability to any fighter that was currently in the Japanese Army or Navy, but then it was revealed that the Nakajima factory had been working on a low-wing monoplane that turned out to be even faster. Concerned that the production contract would be lost, Kawasaki produced a third prototype that was fitted with a three-bladed metal propeller, which increased the aircraft's speed considerably. In addition to this, all the rivets in the metal section of the aircraft were fitted with flush-head rivets, giving the aircraft a streamlined look.

Although it was still slower than the Nakajima Ki-11, the speed difference had been decreased, and when it was discovered that the Ki-10-I was markedly better in manoeuvrability, a large production contact was awarded to Kawasaki. In an effort to improve the stability of the aircraft, the wings were lengthened, as was the fuselage. The Ki-10-II, as it was called, went into production at the beginning of 1936. This was of an all-metal construction with a light alloy skin and fabric-covered control surfaces.

A number of modifications took place, and after the first 200 Ki-10-IIs had left the production line, the radiator was moved from beneath the engine cowling to between the redesigned cantilever undercarriage, which now had wheel coverings to reduce the drag.

The Kawasaki Ki-10-II was assigned to fighter units in Formosa, Korea and Manchukuo and saw action in Manchuria and China during the Second Sino-Japanese War. By the time the Pacific War had started the Ki-10 had been re-assigned to training and communication duties, although a small number were encountered over China. A total of 588 Ki-10s were produced between 1935 and 1937.

NEW CARRIER MONOPLANE AIRCRAFT

The arrival of the Nakajima B5N heralded the last of the biplane fighters. The JNAF issued a specification for a single-engined, monoplane carrier attack bomber that had to have a wingspan of less than 52 ft 5½ in. (16 m) fitted with a folding mechanism that would reduce the span to 24 ft 7¼ in. (7.5 m), a fuselage length of not more that 33 ft 9½ in. (10.3 m), carry a bomb load or torpedo of 1,764 lb (800 kg), a four-hour endurance, but with additional fuel tanks for seven hours, and which could carry a crew of three. The specification had been worked out to enable the aircraft, when in a folded configuration, to sit comfortably on the standard elevator of an aircraft carrier.

The Nakajima design team, under Katsuji Nakamura, set to work and produced what was to be designated the Nakajima B5N1 Navy Type 97 Carrier Attack Bomber Model 1 (B5N1). The aircraft incorporated a folding-wing design with the hinge points arranged in such a way that, when the wings folded over the fuselage, the wingtips overlapped each other in one smooth motion. The undercarriage was hydraulically operated and, together with a number of other technical innovations including the Fowler flaps, was completed in December 1936.

The first test flight of the B5N, powered by a 700-hp Nakajima Hikari 2 9-cylinder air-cooled radial, took place on 3 January 1937 and promptly suffered a failure in the hydraulic system, causing the test to be cancelled for a few days. With the problem resolved, the flight tests took place and exceeded all that was asked. The only problem that concerned the Navy was that the aircraft relied heavily on the hydraulic system, and the number of technical innovations incorporated into the aircraft gave cause for concern with regard to maintenance. A second prototype was produced with the hydraulic wing-folding mechanism replaced by a manual one. The Fowler flaps were replaced with more conventional ones and additional fuel tanks were placed within the wings.

The flight tests were a complete success, and in November 1937 the Navy ordered that the aircraft be put into production. Over the next three years 669 B5N1s were built and saw service on land and aboard aircraft carriers during the Sino-Japanese War. Unusually for aircraft being built around this time, no major modifications were needed to the aircraft, only upgrades to the equipment within the aircraft.

It wasn't until 1939 that a new variant appeared, the B5N2 powered by a 1,000-hp Nakajima Sakae II 14-cylinder radial engine. This had come about because of the improved fighters that the Chinese had obtained from the West. Because the new engine was of a smaller diameter, the designers decided to reduce the size of the engine cowling, giving the pilot an improved view, reducing the drag and improving the cooling.

The B5N2 was still in service in December 1941, and was one of the attack aircraft concerned in the attack on Pearl Harbor. The land-based B5N2s, which had been given the identifying name of 'Kate' by the Allies and had taken part in the Solomons and Philippines Campaigns, found themselves up against far superior Allied aircraft and suffered tremendous losses. They were relegated to second-line units and were used for maritime reconnaissance and anti-submarine roles because of their long endurance capability. A small number of the aircraft were fitted with radar, with the aerials fitted along the sides of the fuselage and along the leading edges of the wings. One variant that appeared in 1943 was the B2N1-K, an advanced trainer that was also used for towing the Chikara Special Training Gliders.

Between 1937 and 1943, the Nakajima, Aichi and the Dai-Juichi companies built a total of 1,149 B5Ns.

While Nakajima was developing the B5N, the Mitsubishi company was working on a similar aircraft. This was designed as a carrier attack bomber with folding wings for flight deck storage. The Mitsubishi Navy Type 97 Carrier Attack Bomber Model 2 (B5M1), as it was designated, was an all-metal, low-wing monoplane that carried a crew of three in tandem. It had a fixed undercarriage with large spats covering the wheels. Initial tests showed its performance not to be as good as other aircraft of its type, but despite this it was put into production late in 1937.

The arrival of the superior B5N1 caused production of the B5M1 to be halted after only 125 had been built. Their role was relegated to being used from land bases in Southeast Asia and China for a short time. They were then used as target tugs, trainers and, towards the end of the war, suicide bombers.

FAST RECONNAISSANCE

The need for a fast reconnaissance aircraft was high on the agenda of the Koku Hombu, and it issued a specification to Mitsubishi that called for a two-seat reconnaissance aircraft that had a top speed of 280 mph (450 km/h), an endurance of one hour at a distance of 250 km from base and a maximum weight of less than 5,291 lb (2,400 kg). Armament was to consist of a single flexible 7.7 mm machine gun mounted in the rear of the cockpit, and it was to be fitted with the latest light aerial camera and with Hi-4 radio equipment.

What resulted was a low-wing, cantilever monoplane, powered by a 750-hp Nakajima Ha-8, 9-cylinder, air-cooled, radial engine, and a fixed undercarriage with spatted wheels. The prototype flew for the first time in May 1936 and exceeded all the performance figures required. Forward visibility from the cockpit, however, left a lot to be desired, especially during landing and take-offs, but once in the air it had excellent handling qualities and manoeuvrability and was enthusiastically received by the pilots.

An endorsement of the aircraft came when one Japan's leading newspapers, the *Asahi Shimbun*, persuaded the Army to let it buy the second prototype and fly it to England for the coronation of King George VI. The aircraft, flown by Masaaki Iinuma, pilot, and Kenji Tsukagoshi, mechanic/navigator, completed the flight from Japan to England in ninety-four hours, seventeen minutes and twenty-three seconds, establishing a world record. More importantly, it proved to the Army the reliability of the aircraft over long distances and highlighted its use as a reconnaissance machine.

The Mitsubishi Ki-15-I Army Type 97 Reconnaissance Model 2, as it was officially designated, was one of the first aircraft to be involved in the second of the Sino-Japanese Wars. Because of its speed of almost 300 mph, it was considerably faster than any of the collection of Chinese aircraft, which consisted of Curtis Hawks, Gloster Gladiators and Russian Polikarpov I-15 and I-16s. This enabled it to fly deeper into Chinese territory from the base in Manchuria, and by doing so was able to keep tabs on almost all of the Chinese ground troop movements.

With the success of the Ki-15-I, a second version appeared, powered by a Mitsubishi Ha-26-I 900-hp, 14-cylinder, air-cooled radial engine and designated the Ki-15-II. The success of the Ki-15-I had attracted the attention of the JNAF, who placed an order for twenty of the aircraft, but fitted with their own radio and aerial camera equipment. Designated the Navy Type 98 Reconnaissance Plane Model I, or C5M1, it was powered by a 875-hp Mitsubishi Zuisei 12, 14-cylinder, air-cooled radial engine. Later, in 1940, a further thirty were ordered and built, but were powered by a 950-hp Nakajima Sakae 12, 14-cylinder, air-cooled radial and were known as C5M2s.

The Army ordered a faster version known as the Ki-15-III and two prototypes were built, powered by the latest Mitsubishi 1,050-hp Ha-102, giving it a top speed of 329 mph (530 km/h). They were about to be flight tested when the Mitsubishi Ki-46 suddenly made its appearance and its performance figures exceeded those of the Ki-15-II, so production was not even started.

At the beginning of the Pacific War, the Army's Ki-15-II, codenamed 'Babs' by the Allies, and the Navy's C5M2 were both operational, and it was one of the Navy's reconnaissance C5M2s that spotted the two British battleships, HMS *Prince of Wales* and HMS *Repulse*, both of which were sunk as a result. For the first year of the war both versions of the aircraft performed their duties well, but as Allied fighters became faster and more deadly, they started to suffer mounting losses. They were withdrawn from active service in 1943 and used for training and communication, and, at the very end of the war, for kamikaze attacks. The Mitsubishi factory between 1936 and 1940 built a total of 489 Ki-15 and C5M aircraft.

At the same time as the Ki-15 made its appearance, the Japanese Army continued to upgrade their aircraft. They had increased their heavy bomber requirement and now looked toward replacing the Kawasaki Ki-3 bomber. In May 1936, with this in mind, they issued specifications for a new two-seat light bomber to the Kawasaki and Mitsubishi companies, and asked them to build two prototypes before the end of the year. The following specifications were laid down by the Army: a maximum speed of 248 mph (400 km/h); a climb rate of 1,500 feet per minute; an operating altitude between 6,560 ft and 13,125 ft (2,000 m and 4,000 m); the ability to carry a maximum bomb load of 992 lb (450 kg); one fixed forward- and one flexible rear-firing 7.7 mm machine gun; the ability to perform dives at an angle of 60 degrees; and a weight not exceeding 7,275 lb (3,200 kg).

The Mitsubishi prototype was initially designed with a retractable undercarriage, but it was decided that the extra speed achieved would be offset by the additional weight and maintenance problems brought about by the complex retraction mechanism. So it was decided to have a fixed undercarriage with spatted main wheels. So that a fuselage bomb bay could be installed, a mid-mounted wing with landing flaps was fitted.

On 28 February 1937, the Mitsubishi Ki-30 Army Type 97 Light Bomber, as it was designated, powered by an 825-hp Mitsubishi Ha-6 14-cylinder, air-cooled radial engine, made its maiden flight. The second prototype, powered by a Nakajima Ha-5 850-hp air-cooled radial engine, made its first flight later the same month. Both aircraft exceeded the requirements laid down by the Army and orders were placed for sixteen aircraft for service trials. There were a number of minor modifications made to these aircraft before the Ki-30 went into full production.

The first Ki-30s off the production line were assigned to the Sixth and Sixteenth *Sentais* and began their operational careers against Chinese forces during the Sino-Japanese War. They became the Army's most reliable aircraft and achieved much success; however, the opposition was almost non-existent, so the aircraft was never really tested against any aircraft of equal ability. This was to become apparent much later during the Pacific War when the Ki-30, codenamed 'Ann' by the Allies, went up against Grumman Wildcats and Hellcats and suffered heavy losses.

They were removed from front-line action and used for crew training, and like many other obsolete Japanese aircraft, were used for kamikaze attacks at the very end.

The relative success of the Ki-30 was instrumental in creating a smaller version of the aircraft that could be used as a ground attack aircraft. The idea came from Captain Yuzo

Fujita, one of Japan's top pilots, who, having flown the Ki-30, realised the potential. The Koku Hombu issued a specification to Mitsubishi that called for a two-seat ground attack aircraft with a maximum speed of 261 mph (420 km/h); an ability to carry a maximum weight of 5,952 lb (2,700 kg); a bomb load of twelve 33 lb (15 kg) bombs or four 110 lb (50 kg) bombs; two fixed forward-firing 7.7 mm machine guns and one flexible rear-firing 7.7mm machine gun.

The design team got to work and produced an aircraft that looked very similar to the Ki-30, but had a much shorter wingspan and was lower, which reduced the length of the fixed undercarriage. Like the Ki-30, it too had spatted wheels. In each wing was mounted a 7.7 mm machine gun and a similar rearward-firing flexible one was mounted in the rear cockpit. The fuselage was also shorter, as was the cockpit area, which had a limited number of flight controls and instruments installed in its rear section. The Mitsubishi Ki-51, as it was called, was powered by a 900-hp Mitsubishi Ha-26-II, 14-cylinder, air-cooled radial engine, which gave the aircraft a maximum speed of 263 mph.

In June 1939 the first of the prototypes appeared, followed two months later by the second. After tests, eleven aircraft were built for service trials and they were delivered to the Army between September and December the same year. A small number of modifications were made during this period, including the fitting of leading-edge slats, which improved the handling characteristics at low speed. Beneath the engine and the cockpit area 6 mm armour plating was installed, one of the few times that armour-plating to protect the crew was fitted in Japanese aircraft.

After extensive trials the Mitsubishi Ki-51 was put into production and only two major modifications were ever made to the aircraft. These were the installation of two wing leading-edge fuel tanks containing 15 gallons (68 litres) each and the replacing of the two wing-mounted 7.7 mm machine guns by two 12.7 mm machine guns.

The first of the production aircraft saw action during the Sino-Japanese War and was well liked by the crews because it was an easy aircraft to fly and very manoeuvrable. The maintenance staff also approved it because it could easily be maintained on front-line airfields.

Because of the relative success of the Ki-51 in China, one of the models was taken away and a retractable undercarriage fitted, together with two wing-mounted 20 mm cannons. A more powerful engine was fitted, but there were serious mechanical problems and any ideas about modifying the aircraft were shelved.

During the Pacific War the Ki-51 performed well in the early months, but as faster and well-armed Allied fighter aircraft entered the war it became an easy target for the new fighters and was withdrawn from front-line duties. Like many other Japanese aircraft, towards the end of the war the Ki-51 was reduced to flying kamikaze missions. Between June 1939 and March 1944, the Mitsubishi and Tachikawa Companies built a total of 2,385 Ki-51 aircraft.

THE DEMAND FOR LONG-RANGE AIRCRAFT

With the fighting in China becoming increasingly bitter and resources being stretched to almost breaking point, senior air officers demanded the need for long-range fighters. This had come about because of the tactics the Chinese were using: keeping their fighters out of the range of the Navy Type 96 Fighters that were currently operating in the areas. Consequently, this meant that the Navy bombers had to go deep into enemy territory without a fighter escort and they were suffering heavy losses. After a great deal of consultation with the senior air officers, a specification was issued to both the Nakajima and Mitsubishi companies for the design and development of a long-range fighter.

The specifications called for a three-seat, twin-engined aircraft that had a maximum speed of 322 mph (280 kt), a maximum range of 2,302 miles (2,000 nautical miles) and armament that included a forward-firing 20 mm cannon and 7.7 mm machine guns and a flexible rear-firing 7.7 mm machine gun.

The Nakajima company produced one of the most technically innovative aircraft of the day. The low-winged monoplane was powered by two 1,130-hp Nakajima Sakae 21 and 22 radial engines, which rotated the propellers in opposite directions to offset the torque generated by them. The armament consisted of a forward-firing 20 mm cannon and two 7.7 mm machine guns mounted in the nose, and two remotely controlled barbettes with 7.7 mm machine guns mounted in tandem on top of the fuselage behind the cockpit. These were hydraulically operated, as were the undercarriage and the cowling flaps.

The first flight trials were plagued with teething problems concerning the very complex hydraulic system that affected the landing gear, the remotely controlled barbettes that the pilot found almost impossible to aim with any accuracy, and the contra-rotating propellers. The aircraft was also considerably overweight, which caused handling difficulties for the pilot. During tests with the Navy the J1N1, as it was called, was pitted against a single-engined Mitsubishi A6M2. In every aspect bar one (the range), it was inferior and the trials were a disaster. The aircraft was returned to the Nakajima factory and re-designed as a land-based, long-range reconnaissance aircraft.

By removing the two barbettes together with their guns and the remainder of the armament, and by reducing the internal fuel capacity, the weight problem was resolved. In an effort to retain the range, drop tanks were fitted beneath the wings. The cockpit area was re-designed to accommodate the pilot and the radio operator/gunner in one section, while the navigator/observer was located in a separate cockpit behind the leading edge. Given the designation of Nakajima J1N1-C (Gekko), the prototype was sent to the Navy for evaluation. This time it was accepted without difficulty and put into production.

Initially production was slow, mainly because the need for reconnaissance aircraft was way down the list of priorities for the Navy, but a small number were fitted with a spherical turret behind the pilot's seat to allow the radio operator/gunner to use it as a defensive weapon. Then the CO of the 251st *Kokutai* put forward the idea to use the J1N1-C as a night fighter, so all the equipment in the observer's cockpit was removed, and two fixed 20 mm cannons were installed at an angle that enabled them to be fired forward and upward at an angle of 30 degrees.

A few weeks later, one of the night fighters intercepted and destroyed two B-24 Liberators. The news of this filtered back to the Navy's High Command, who saw the advantages in having a night fighter and ordered the Nakajima company to build a version based on the one converted by the 251st *Kokutai*. The prototype J1N1-S Gecko Model 11 appeared some months later and production began in August 1943. Between August 1943 and March 1944, 183 night fighters were built. Only fifty-four of the J1N1-C models were built, which shows the importance given to the night fighter.

Production of the J1N1-C was halted in December 1944, by which time a further 240 J1N1-S Geckos had been produced. As the war began to end, what was left of the Nakijima J1N1-S and Rs were used in kamikaze attacks carrying two 551 lb (250 kg) bombs.

Earlier a Nakajima Army Type 92 (Ki-20) Heavy Bomber, powered by four 800-hp Junkers L88 12-cylinder engines, had been built, but was rejected by the Army as being too unstable. This was the largest aircraft ever built by the Japanese. It had a wingspan of 144 ft 4¼ in. (44 m), a length of 76 ft 1½ in. (23.2 m) and a height of 23 ft (7 m). It was capable of carrying a normal bomb load of 4,490 lb (2,000 kg) and a maximum bomb load of 11,023 lb (5,000 kg) and was armed with twin 7.7 mm machine guns in the nose, twin 7.7 mm machine guns in two upper wing turrets, a single 7.7 mm machine gun in both lower wing turrets, and one 20 mm cannon mounted on top of the fuselage.

The design of the aircraft was based on the Junkers G.38 passenger aircraft after two of Mitsubishi's top designers, Nobushiro Nakata and Kyonosuke Ohki, had been sent to Germany in 1928 to study the design and construction of the aircraft. In 1930, the company purchased the necessary machine tools and jigs, and after obtaining a licence from Junkers, started to build the aircraft in Japan. The first two aircraft were built using parts made in Germany, but the remaining four aircraft were built entirely out of Japanese-made parts. In total six of the aircraft were built, but because of the various problems, were used as flying test beds for various engines. One of the major problems that faced the Army was the limited number of airfields capable of accommodating this lumbering giant. During the Manchurian campaign the use of the aircraft was strictly for research and it was never flown in combat.

Not to be discouraged, the Army then initiated a design contest for a new twin-engined heavy bomber to replace the Type 92. Two companies participated, Mitsubishi and Nakajima, who by now were two of the premier manufacturers of aircraft in Japan. Competition was fierce between the two companies and in 1935 the first of the two prototypes appeared, the Nakajima Ki-19 and the Mitsubishi Ki-21. The aircraft chosen by the Army was the Mitsubishi Ki-21, and this was to become one of the Army's principal bombers from its conception until Japan surrendered in 1945.

It was a radical design, inasmuch as it was a twin-engined, all-metal monoplane with a wingspan of 73 ft 10 in. (22.5 m), a length of 52 ft 6 in. (16 m) and a height of 14 ft 5 in. (4.35 m). It was powered by two 825-hp Mitsubishi HA-6, 14-cylinder air-cooled, radial engines, giving it a maximum speed of 268 mph (496 kts). The first batch of production models were beset with teething problems, but once these had been sorted out the Mitsubishi Army Type 97 (Ki-21) Heavy Bomber, as it was designated, became one of the most reliable of the Army's bombers. A total of 2,064 of the aircraft were built, one of the largest numbers of aircraft ever produced by Japanese aircraft manufacturers.

The Kawasaki prototype Ki-32 was produced at the same time, but this was a single-engined light bomber. Eight prototypes were produced and all suffered problems with the engine that caused the engine nacelle to be redesigned and the crankshaft to be strengthened. During trials against the Mitsubishi Ki-30, the Ki-32 proved to be the better aircraft, but the engine problems caused the Army to go for the Mitsubishi Ki-30 because of its reliable engine. The Army had a full-scale war on their hands in Northern China and Manchuria and aircraft reliability was paramount.

In July 1938, with the engine problems resolved, the Kawasaki Army Type 98 (Ki-32) was put into production and a total of 846 were built between July 1938 and May 1940. The Type 98 light bomber was an immediate hit with the crews and within months was in action in the second part of the Sino-Japanese conflict.

With the bomber part of their Air Wing seemingly settled the Army looked at its fighter aircraft, which were becoming obsolete as technology gathered pace. They initiated a design competition between the major companies, of which only three participated: Kawasaki, Nakajima and Mitsubishi. Kawasaki produced a biplane powered by an 850-hp Ha-9 engine with the designation Ki-5. With the monoplane becoming more and more prevalent, the introduction of a biplane seemed to be a retrograde step, but it had exceptional manoeuvrability qualities.

Designed by Takeo Doi, the prototype had an unequal wingspan and was initially fitted with a two-bladed wooden propeller. Designated the Ki-5, it was rejected by the Army as being unstable. After a number of major modifications, which included changing the propeller to a three-bladed metal version and using flush-headed rivets in the construction of the fuselage, the improvement in the performance in the Ki-10, as it was now designated, was markedly superior to that of the Ki-5. A second prototype was built and both aircraft were submitted to the Army for test and valuation.

A third and fourth prototype were later submitted, and although slower than the Nakajima Ki-11, the Kawasaki Army Type 95 (Ki-10) was put into production and in the next two years over 300 of the aircraft were built.

The need for another faster fighter arose with the outbreak of hostilities between Japan and China for the third time. It was the Navy that had to feel the brunt of the war during this period, as the Army's Air Wing was in the process of re-equipping its Air Wings. The Army instigated a competition for a new fighter to replace the Kawasaki Type 92 Fighter that was currently in use but becoming rapidly obsolete. Kawasaki had earlier submitted the Ki-10, which had been accepted and was now in production, while Nakajima entered its Ki-11, a wire-braced, low-wing monoplane fitted with wheel spats, with an open cockpit and based on the design of the American Boeing P26. Powered by a 700-hp Nakajima Ha-8 engine, which was lower rated than the Kawasaki Ki-10, the Ki-11 was considerably faster. However it was rejected on the grounds that it was less manoeuvrable than the Ki-10, but Nakajima used the design to progress on to what was to be designated the Type PE.

In 1937, the Army put forward a proposal for a twin-engined, long-range fighter. Three companies, Kawasaki, Mitsubishi and Nakajima, submitted designs to the Koku Hombu (Air Headquarters): the Ki-38, Ki-39 and Ki-37 respectively. After negotiations both Mitsubishi and Nakajima withdrew because of long-term commitments, leaving Kawasaki's Ki-38 to be considered. For reasons known only to them, the Koku Hombu shelved the idea. One year later they approached Kawasaki to initiate work on the development of a two-seat, twin-engined fighter with specifications that gave it a maximum speed of 335 mph (540 km/hr), an operating ceiling of between 6,560 ft and 16,405 ft (2,000 m and 5,000 m), and an endurance of at least 4 hours. It was to be equipped with two forward-firing guns and one flexible rear-firing gun.

Powered by two wing-mounted 850-hp Nakajima Ha-20b engines housed in large nacelles with an exhaust collector ring in front, the first prototype Ki-45, as it was known, was sent for test and evaluation. The initial response was not favourable because of excessive drag caused by the large engine nacelles and problems with the manually retractable undercarriage. The second prototype had closer fitting engine cowls and propeller spinners fitted. It required a third prototype to resolve the problems with the manual retraction of the undercarriage, and that was achieved by installing an electrically-operated retraction mechanism.

But still there were problems regarding the speed of the aircraft, and Kawasaki was ordered to install two 1,000-hp Nakajima Ha-25 14-cylinder radials with smaller spinners fitted to the propellers onto another of the prototypes, to see if that would improve the performance. It did, and the modification was considered to be a success. Whilst this was going on Kawasaki's chief designer, Takeo Doi, had been listening to the criticisms and observing the modifications, and he decided to redesign the aircraft taking into account the modifications that had proved to be successful. The result was the Ki-45 KAI, which had a slimmer fuselage and redesigned fabric-covered tail surfaces, straight tapered wings that had an increased span and wing area, smaller diameter engine nacelles mounted lower down on the wing, and new machine guns. The first of the prototypes was sent for evaluation and was accepted and at the end of 1941 the aircraft was designated the Army Type 2 Two-seat Fighter Model A (Ki-45 KAIa) Toryu ('Dragon Killer'). A total of 1,701 of this model were built at the Kawasaki plants of Gifu and Akashi.

A later model was designed specifically as a night-fighter, the Ki-45 KAIc. The forward armament was removed from the nose section, which was designed to accommodate a radar section, while two obliquely mounted upward-firing 20 mm cannons in the centre section of the fuselage and a 37 mm semi-automatic cannon were fitted in the ventral tunnel.

Also in 1937, although the main interest for the Army and the Navy were in either fighters or bombers, the Army took a long, hard look at developing a transport aircraft. The Nakajima company had acquired the manufacturing rights of the Douglas DC-2 in 1935 and had built a small number of these aircraft, known as Nakajima AT-1 & 2, for Japan Air Lines and the Manchurian airlines. The Army saw a use for these aircraft as communications and paratrooper transport and ordered Nakajima to build them to military specifications. Designated the Nakajima Ki-34 (AT-1) (codenamed 'Thora' by the Allies), 299 of these aircraft were built between 1939 and 1942 and used extensively during these periods. This aircraft was based on the designs of the Douglas DC-2 and Northrop 5A and was powered by two 580-hp Nakajima Kotobuki 2-1 radial engines.

The first prototype was produced in September 1936 and was flown by a variety of test pilots ranging from military to civil. The aircraft performed well and was considered by all as an extremely stable aircraft with good manoeuvrability characteristics. A few minor problems were detected with the engine cooling system and the mechanism that retracted the undercarriage. Both these problems were easily fixed and the aircraft was put into production.

The production models had both engines replaced with two 710-hp Nakajima Kotobuki 41s and were built for both the military and civil markets. The JAAF used the Ki-34 as a paratroop transport and as a communications aircraft. Such was the commitment of Nakajima to other projects that production of the majority of the Ki-34 was shifted to the Tachikawa company, who built 299 of the 351 produced, the remainder being built by Nakajima.

In 1938 some of the Ki-34 AT-Is given to the JNAF, and given the designation Navy Type AT-2 Transport, were used as both a communications and transport aircraft.

NEW LONG-RANGE FLYING BOATS

In 1934, the Navy had issued a specification that called for a large monoplane four-engined flying boat of an all-metal construction with fabric-covered control surfaces, a range of 2,500 nautical miles and a cruising speed of 125 knots. This would make it superior to the American Sikorsky S-42, which was the premier flying boat of its day. Earlier that year, a Japanese team from Kawanishi, led by design engineers Yoshio Hashiguchi and Shizuo Kikahura, had visited the Short Brothers factory in England. With the information obtained they designed a four-engined flying boat that had a parasol wing mounted on inverted V-struts above the fuselage, with parallel supporting wing bracing-struts running from the lower section of the hull to the mid-span of the wing.

Known as the H6K, it was powered by four 840-hp Nakajima Hikari 2 9-cylinder, air-cooled, radial engines. On 14 July 1936, the first prototype lifted into the air and the aircraft performed well both on the water and in the air. In an effort to further improve the water handling characteristics, the forward step on the hull was moved back 1 ft 7¾ in. (50 cm). Further trials were carried out satisfactorily and with those completed, the aircraft was handed over to the Navy for further testing and evaluation.

The tests went well, but the Navy was concerned with the lack of power and suggested that the engines be replaced with more powerful ones. Defensive armament consisted of one flexible 7.7 mm machine gun mounted in the open bow position, one 7.7 mm machine gun in a power-operated dorsal turret (the first Japanese aircraft to have one) and one flexible 7.7 mm machine gun in a fixed rear turret. The H6K1 could also carry 1,764 lb (800 kg) torpedoes and 2,205 lb (1,000 kg) of bombs attached to the parallel wing supporting struts.

Two more prototypes were built and subjected to extensive service trials. They differed inasmuch as they had enlarged tail fins, larger ailerons and a redesigned dorsal turret. They were also fitted with four 1,000-hp Mitsubishi Kinsei 43 14-cylinder, air-cooled radial engines and were given the designation of Navy Type 97 Flying Boat Model 1 (H6K1). In January 1938 the H6K1 went into production. Only three were built before the next version, the H6K2. The first two of these were deemed to be prototypes for a transport version, while another eight came off the production line before the next version, a long-range maritime reconnaissance flying boat, the H6K4, appeared. The H6K3 never actually appeared as military aircraft; two prototypes were built as a civilian version but were never put into production, as the need for military aircraft was more pressing at the time.

The H6K4 went into full production in 1940 and a number of modifications were made, including increasing the fuel capacity from 1,708 gallons (7,765 litres) to 2,950 gallons (13,409 litres). The power-operated dorsal turret was removed and replaced with an open dorsal position, two blisters either side of the fuselage containing one hand operated 7.7 mm machine gun were fitted and a 20 mm cannon was fitted in the tail turret. One year later another version appeared powered by four 1,070-hp Mitsubishi Kinsei 46 engines. A total of 127 of these versions were built between 1939 and 1942.

In 1941 the H6K4 was used for bombing missions at Rabul and the Dutch East Indies, but as the Allies began to fight back and enemy fighters became more and more powerful, the lack of armour protection and self-sealing fuel tanks on these flying boats caused a dramatic increase in losses. The H6K4 then resumed its original role as a long-range maritime and patrol aircraft, using its remarkable endurance to good effect. The open bow position was replaced with a housing, still containing a 7.7 mm machine gun, and four 1,300-hp Mitsubishi Kinsei 53 radial engines were fitted, and became known as the H6K5. With the production of the H6K5 flying boats well under way, the Kawanishi company produced a transport version known as the H6K2-L, which was based on the design and specifications of the H6K4. All the armament was removed and the interior changed to accommodate a mail and cargo compartment forward of the flight deck, a galley behind the flight deck, a mid-fuselage section with eight seats or four sleeping bunks, behind this a ten-seating accommodation section and behind that a toilet and cargo compartment. These transports were given the code name 'Tillie' by the Allies, while the reconnaissance flying boats were known as 'Mavis'.

A total of 215 H6Ks in one form or another were built between 1936 and 1942 and this was one of Japan's most successful military aircraft.

The success of the H6K prompted the JNAF to put forward a proposal for another large four-engined maritime reconnaissance to be built. The Allies had now developed their own long-range maritime reconnaissance flying boats, the British Short Sunderland and the American Sikorsky XPBS-1, both of which were capable of covering vast areas of the Pacific Ocean. The Japanese version had to be capable of exceeding all the known requirements of these two aircraft, and so the Kawanishi company was approached with the proposal.

In 1938 the Kawanishi team got to work designing a flying boat that would meet all the stringent requirements laid down by the Navy. Almost two years later the H8K, as it was known, emerged from the factory for its first flight. The H8K was of an all-metal construction with eight small, unprotected fuel tanks in the wings, four either side, and six large tanks in the hull. The tanks in the hull had a carbon dioxide fire extinguisher system fitted inside them, with the additional protection system that if punctured, the fuel inside would drain away into the bilges, where pumps would send it back into the undamaged tanks. As expected in a long-range reconnaissance aircraft, the fuel carried represented almost 30 per cent of the aircraft's maximum take-off weight. Additional weight was added with the extensive armour protection provided for the crew of ten. The stabilising floats were fixed to save weight, although initially they had been designed as retractable floats.

Powered by four 1,530-hp Mitsubishi MK4A Kasei 11 14-cylinder air-cooled radial engines, the H8K1 had a maximum speed of 269 mph at 16,450 feet. Its defensive armament consisted of one 20 mm cannon mounted in the nose section, one in the dorsal turret, one in the tail turret and one in each of the blisters on either side of the fuselage. Flexible 7.7 mm machine guns were also mounted in the two side hatches and in the ventral positions.

The first test flight threw up a major problem during high-speed taxiing and take-off. It was discovered that the shallow depth of the hull caused water to be thrown into the engines and over the wings. The aircraft was returned to the factory, where the depth of the hull was increased by 1 ft 8 in. (50 cm). The planning of the bottom of the hull was also modified and longitudinal steps were added between the forward section of the hull between the keel and the main chine, which solved the problem.

The aircraft was sent to the Navy for service trials and at the end of 1941 was accepted by the Navy and given the official designation Navy Type 2 Flying Boat Model 11 (H8K1). Two pre-production models were produced and a number of minor modifications made before the aircraft went into full production. These consisted of reducing the number of defensive guns and adding facilities that enabled the flying boat to carry an offensive load, which consisted eight 551 lb (250 kg) or sixteen 132 lb (60 kg) bombs, or two 1,764 lb (800 kg) torpedoes.

On the night of the 4/5 March 1942, the first offensive sortie concerning the H8K1 took place. Two of the aircraft belonging to the Yokohama Kokutai, based on the Wotje Atoll in the Marshall Islands, carried out a bombing attack on Oahu Island, part of the Hawaiian Islands. On approaching their target, the crew experienced very heavy cloud cover and their bombs caused almost no damage to property or inhabitants. On its return to base, more problems faced them when the refuelling submarine that carried the fuel for the aircraft surfaced at the French Frigate Shoals only to find it occupied by US forces.

The H8K1's role as a bomber was relatively short-lived, but its role as a long-range maritime reconnaissance aircraft, for which it was originally intended, proved to be extremely successful. Its high speed and good defensive armament showed it to be an aircraft that could defend itself more than adequately.

The arrival of the H8K2, powered by four 1,850-hp Mitsubishi MK4Q Kasei 22 engines fitted with water injection, increased the speed of the aircraft to 290 mph. The performance was so markedly improved that only sixteen of the H8K1s were built before the H8K2 with the new engines appeared. There was also some slight modification to the vertical tail surfaces, but other than that the H8K2 was identical to the H8K1 in appearance. The increased engine power allowed for a greater take-off weight, which meant that the aircraft could carry more fuel, therefore increasing its range and endurance. Codenamed 'Emily' by the Allies, the H8K was one of the fastest and most heavily-armed reconnaissance aircraft in the Pacific War theatre and was considered by many Allied fighter pilots to be one of the most difficult to shoot down.

In 1943, the need for large transport aircraft prompted the Navy to take the H8K1 prototype and convert it into a troop/staff transport. The hull had been deepened, which enabled two separate decks to be built: one from the nose section to the rear hull step, the other from the centre section of the wing to the rear of the fuselage. With the exception of the 7.7 mm machine guns in the nose and rear turrets, all other armament was removed. The hull fuel tanks were removed to accommodate the lower deck, and this reduced the fuel capacity of the aircraft and ultimately its range and endurance.

Known as the Kawanishi Navy Type Transport Flying Boat *Seiku* (H8K2-L), thirty-six of the aircraft were built and used as troop transports throughout the Pacific War. Two of the H8K3 flying boats were modified by taking away the two side blisters and replacing them with sliding hatches. Also, the fixed dorsal turret was replaced with a retractable dorsal turret and the fixed stabilising floats were replaced with retractable ones.

A total of 167 H8Ks in various models were built between 1940 and 1945, but the Japanese Navy halted production in the last year because of the desperate need for more fighters and interceptors. It was without doubt one of the Japanese Navy's most successful water-based combat aircraft of the Pacific War.

One of the most advanced seaplanes of its type appeared in 1941, when the JNAF approached the Kawanishi company with a proposal for a single-engined, high-speed, two-seat reconnaissance seaplane that could out-perform any of the Allied land-based fighters. In addition to this, the company was given almost carte blanche freedom in the design and requirements. This was virtually unheard of, as nearly all design proposals and requirements were accompanied by stringent specifications.

The design produced was of a floatplane of an all-metal construction, which had a single central float that was cantilever pylon mounted, and was attached to the fuselage by two pins. In an emergency the forward pin could be extracted, releasing the float. Two rubberised-fabric retractable stabilising floats with metal planing bottoms were fitted to the tips of the wings, which were inflated when the floats were extended. Armament consisted of a single flexible 7.7 mm machine gun mounted in the observer's position.

The floatplane was powered by a 1,500-hp Mitsubishi MK4D Kasei 14 air-cooled radial engine that turned two two-bladed, contra-rotating propellers. The first test flight took place on 5 December 1941, and problems were encountered with both the contra-rotating pitch control system and the retractable stabilising floats. During one test flight, the stabilising floats became inoperable because of flap failure and considerable damage was caused to the aircraft when it touched down on the water. The stabilising float retracting system was to plague the E15K throughout its relatively short life.

Despite the problems, the aircraft went into production and was given the designation of Navy High-Speed Reconnaissance Seaplane (E15K1). The retractable stabilising floats were replaced with fixed floats that were attached to the wings with cantilever struts. The combat experience of the E15K1 was non-existent: the only known sortie was when six of the aircraft were sent to Palau for combat evaluation and were attacked by Allied fighters. They attempted to jettison their ventral float to give them more speed and manoeuvrability, but none of the six aircraft were able to do so and all were shot down. A total of fifteen E15K1 seaplanes were built: six prototypes and nine production models.

THE LEGENDARY ZERO

The number of new types of experimental aircraft being designed and built increased almost monthly, but hardly any got past the design or prototype stage. Improved versions of existing fighters were put forward, but in the main were not much better than the existing ones. That was until the JNAF put forward a proposal in May 1937, and approached two of Japan's leading aircraft manufacturers, Mitsubishi and Nakajima, with specifications for a new fighter. The requirements of the Navy were so demanding that Nakajima took one look and bowed out, saying that other commitments would have to take priority. This left only Mitsubishi, and the whole project was placed in the hands of their chief design engineer, Jiro Horikoshi.

The specifications laid down were as follows:

It must have:

1. A wingspan less than 39 ft 4 in. (12 m)

2. A maximum speed that would exceed 310 mph (270 knts) in level flight

3. A climb rate of 3,000 ft per minute

4. A flight endurance of between 1 and 1½ hours

5. A range of 1,010 nautical miles (1,870 km) with normal load and 1,685 nautical miles (3,110 km) with external fuel tanks

6. An ability to take off from the deck of an aircraft carrier

7. A landing speed of less than 66 mph (58 knots)

8. Armament consisting of two Oerlikon Type 99 20 mm machine guns and two 7.7 mm machine guns

[Plus a number of other smaller requirements]

The aircraft had to be capable of intercepting and destroying enemy bombers, to serve as escort fighters with their own bombers and to have a combat performance greater than the enemy's fighters. In short, the requirements were for the design and development of what could be the best fighter aircraft in the world – that is, if they could be met. Horikoshi was faced with probably the most demanding problem he would ever have to meet, so he started by selecting the engine that would power this revolutionary aircraft. By a process of elimination he chose the engine that would suit the best, the 870-hp Mitsubishi Zuisei 13. There were a couple of other, more powerful engines, but they were much heavier, and that would require a heavier airframe, which in turn would need more fuel, therefore requiring larger fuel tanks, increasing the weight even more.

One of the problems Horikoshi also faced was that he was having to design a fighter known as the Experimental 12-Shi Carrier-Based Fighter, which was going to face enemy fighters that had engines almost twice the power of his. The Navy were well aware of this and insisted that the engine that should be used was the Nakajima 950-hp Sakae 12, and that the design should also incorporate retractable landing gear. Horikoshi had originally decided to have a streamlined, fixed undercarriage, but the Navy's insistence was, of course, going to increase the weight, so savings had to be made in other areas, and one of these was in the design and construction of the fuselage and the wing structure.

The wing spar was made in a continuous piece and by doing this the wing attachment points were eliminated, making the cockpit area an integral part of the wing. The wing spar is generally accepted as being the heaviest single piece of the structure, and to further reduce the weight this was made of a newly developed alloy. Known as Super Ultra Duralumin (SUD), the alloy was 40 per cent stronger than any other known alloy. It was discovered later that, although it served its purpose at the time, it suffered from strain lines when being rolled during its forming process, causing inner granular corrosion to develop after a few years.

The designers also built a permanent warp into the wing so that the angle of the wing setting, known as the incidence, which was set at the wing root in degrees, decreased as it moved towards the wing tip. The warp was barely noticeable unless looked at down the wing from the wing tip, but it was sufficient to help prevent stalling

By separating the rear section of the fuselage from just behind the cockpit for ease of transportation and maintenance, another saving in weight was achieved. Among the other design features was the all-round vision teardrop-shaped canopy. Up to this point, most designers favoured building the canopy to the raised section of the fuselage behind the pilot's headrest, thus restricting his rearward vision. This new concept allowed the pilot almost 360 degrees of vision, giving him a distinct advantage during a dogfight.

With the design of the aircraft nearly completed, a meeting with the Kaigum Koku Hombu and two of the JNAF test pilots took place to evaluate what had been achieved up to date. Jiro Horikoshi stated how he felt progress was going with regard to the design and construction of the prototype. He was followed by one of the Navy test pilots, Lt-Com. Minoru Genda, who stated that as a carrier-based fighter, the main thing that the fighter had to be able to do was to engage the enemy in close-range combat, and to do that he would like to see the heavyweight cannons replaced with lighter machine guns, which would improve the aircraft's manoeuvrability. This was hotly disputed by the other test pilot, Lt-Com. Takeo Shibata, who made the point that the JNAF fighter aircraft were already superior in close-combat fighting and that more attention should be paid to increase the speed and range of the aircraft. He highlighted this need by saying that the air battles over China being experienced by the Japanese bombers were taking place without the protection of the fighters as the battles were beyond their range.

Shibata also made the point that pilots needed more combat training, saying that the maximum speed of an aircraft was strictly limited by the power of the engine and the design and construction of the aircraft, which the pilot has no control over. However with training, a pilot's skill could compensate for this and give him superiority over his opponent.

The Mitsubishi design teams decided to take another look at the list of specifications laid down by the Navy to see if some compromise could be agreed upon. The insistence on the larger engine and the increased range demanded were going to put serious problems of additional weight on the aircraft. Some Navy pilots wanted a high-speed, fast climbing fighter with a long range. Others wanted a light, fast, manoeuvrable fighter that could easily be controlled in a dogfight. To cover all aspects would require two different types of fighter, and this was not what the Navy wanted; they wanted one that would encompass all these requirements and meet all the needs of the pilots.

Despite all the pressures, Jiro Horikoshi and his team built the prototype, which was on schedule for its first flight in April 1939. But first they had to get the aircraft from the factory to the nearest airfield. Unlike Western aircraft manufacturers, nearly all Japanese companies grew up in heavy industrial areas that only used railways or canals to transport their goods. There were a couple of aircraft manufacturers who were lucky enough to have enough open space beside them to be used as an airfield, but Mitsubishi was not one of them.

Mitsubishi had an airfield at Kagamigahara, thirty miles to the north of Nagoya. Because of the state of the roads, or in some cases the lack of roads, the aircraft was disassembled and packed into large shipping crates. The crates were then placed on two oxen-driven carts, and on 19 March 1939 it was transported to the airfield. Oxen-drawn carts were used because of the dreadful conditions of the roads and paths: using motor trucks would have caused the crates to be jolted heavily, possibly causing serious damage to the aircraft.

The journey began at 7 o'clock in the evening, and the two carts lumbered along through the night and into the following day. By lunchtime the following day the two carts had arrived and the crates were taken into the hangar. The aircraft was re-assembled and prepared it for its maiden test flight.

The morning of 5 April saw the aircraft ready, but because the airfield was also used for Army flight training, it wasn't until late afternoon that it was clear. Mitsubishi's chief test pilot, Katzumo Shima, was to carry out the flight test. The first tests carried out were to test the fighter's ground-handling capabilities and after a series of rapid accelerations, turns and emergency stops, he took the aircraft back to the hangar. The engineers quickly swarmed over the aircraft to question him about the aircraft, and found that what concerned him was the poor braking. Adjustments to the brakes were made and Shima took the aircraft out again for similar tests; this time he seemed satisfied and taxied the aircraft to the end of the runway.

With all eyes on the aircraft, Katzumo Shima opened the throttles and the aircraft roared down the airfield. Seconds later it was airborne and rose to about thirty feet off the ground before settling back down again. The first flight of the Mitsubishi 12-Shi Carrier Based Fighter (A6M2 Reisen (Zero, a.k.a. 'Zeke')) had taken place. Taxiing back to the hangar, Shima clambered out of the cockpit with a broad grin on his face that said it all. His only reservations were the brakes, but other than that he said the aircraft handled very well indeed. The poor braking system was to dog the aircraft throughout its operational life.

Over the next twelve days, Katzumo Shima and his deputy Harumi Aratani made numerous test flights. All the initial flights were made at low level, and with the undercarriage extended. Both the pilots made the observation that there was undue vibration throughout the flights, and after extensive tests it was decided that it was the two-bladed propeller was causing the resonance between the engine and the airframe. The propeller was replaced by a three-bladed version and immediately the vibration disappeared.

With the initial flight tests over, the testing of the aircraft was taken to a new level and it was then found that at high speeds the elevators became overly sensitive. One suggestion put forward was to manually adjust the sensitivity by using a variable control linkage, which was relative to the air speed. This was acceptable under normal flying conditions, but in a combat situation where speeds are constantly changing, the pilot would not have time to make any adjustments. Horikoshi had anticipated that this could be a problem during the design stage and had been working on a possible solution. The problem lay in the standard cable control system, so setting the standards aside, Horikoshi changed the connecting control cables to ones with finer strands, and replaced the torque tubes with more flexible ones. It worked!

One month later, the Kaigum Koku Hombu informed Horikoshi that he was to install the Nakajima Sakae 12 engine. This meant that there had to be some small design changes, resulting in a slightly modified tail. With these completed and flight tests carried out successfully, the Mitsubishi A6M1 Reisen was accepted by the Navy and put into production. Japan now had the world's most modern fighter aircraft. The name 'Zeke' had been attributed to the aircraft by the Allies.

On 14 September 1939, the Mitsubishi A6M1 Reisen was officially accepted by the Navy. With a few modifications, the main one being a correction in the elevator control force, the second prototype was delivered to the Navy.

While other aircraft were being developed, in March 1940 the testing of the A6M1 took a downward turn when one of the production models crashed during dive tests. The pilot, Masumi Okuyama, was carrying out dive tests at the JNAF test field at Oppama; when diving from a height of 5,000 feet, a shrill whine came from the engine, followed by a loud explosion as the aircraft disintegrated in the air. The pilot was seen to come out of the aircraft, followed shortly by his opening parachute. He was then seen to slip out of the parachute harness and plunge into the water from around 1,000 feet – he died instantly.

An investigation followed immediately, and the official explanation was that the accident was caused by excessive fluttering of the elevators caused by the elevator mass-balance arm breaking, resulting in the counter-balance weight shearing off. This excessive fluttering resulted in the engine shaking loose from the airframe, as this was the one weak structural point in the aircraft.

The vast majority of the engineers and pilots did not really accept the explanation, but this was the official line and who was to argue? One year later a second incident, this time with an A6M2 from the aircraft carrier *Kaga*, happened during G-load tests. The pilot, Sub-Lt Yasushi Nikaido, noticed the skin on the wings distorting and starting to buckle during one particularly tight turn. In an effort to find out what was causing the problem, he carried out a series of tight turns and dives. Suddenly the aircraft pitched up violently, causing him to momentarily lose control. When he regained control he discovered both his ailerons were missing and large sections of the skin from the upper sections of the wings had gone. Fortunately he managed to land at a nearby naval base.

The following day an experienced test pilot, Lt Manbei Shimokawa, took off from the aircraft carrier *Akagi* to carry out similar tests in an effort to find the cause. After carrying out a number of dive tests from various heights, he entered a dive from 13,000 feet and pulled out steeply at around 6,500 feet. Suddenly pieces started to fly off, and the aircraft tumbled around in the sky before going into a dive and crashing into the sea. The pilot was killed.

An investigation concluded that the problem had been caused by excessive wing fluttering, and 5 g limits were placed on dive pullouts.

Modifications were made, which included increasing the thickness of the skin on the outer wing, fitting additional external balance weights to the ailerons and installing longitudinal stringers in an effort to increase torsional strength.

With the problem resolved, production began and by the end of 1940 just over 100 A6M1 fighter aircraft had been built. In 1941 the rate of production was increased as the rumblings of war in the Pacific began to become more ominous, and no end to the Sino-Japanese War seemed to be in sight. One Zeke per day was now coming out of the Mitsubishi factory, but at the beginning of 1942, with Japan now at war with America and her allies, the major production of the aircraft was switched to the Nakajima factory, while Mitsubishi concentrated on engineering development.

Without question, the Mitsubishi A6M fighter was Japan's finest aircraft and the one most feared by the Allies. It made its first battle debut in the skies over China, but it was to be a further sixteen months before it made its second appearance, this time at Pearl Harbor.

The appearance of the Zeke at Pearl Harbor stunned the Americans, as they had no idea that the Japanese had developed such a fighter aircraft. In the coming months the only description of the aircraft available was that given by pilots who encountered the fighter. But then, on 3 June 1942, during an attack on Dutch Harbor in the Aleutian Islands, Petty Officer Tadayoshi Koga, flying an A6M2, developed engine trouble. Unable to continue with his mission, he had no alternative than to carry out a wheels-up landing on a large, flat and desolate area on one of the islands. Touching down on a marshy surface, the aircraft flipped onto its back, killing the pilot. Five weeks later, a US Naval scouting party found the aircraft almost intact and set about recovering it.

After weeks of work, the aircraft was recovered almost intact and taken to the US Naval repair depot, NAS North Island, San Diego. The majority of the repair work was required on the tail section, the nose and the canopy. In October 1942, the A6M2 Zeke took to the air once again, only this time in US Naval markings. A variety of American fighter aircraft were pitted against the Zeke in a number of exercises, and assessments were made against each of the aircraft.

At the end of 1940, the Navy had put forward a proposal for a single-seat fighter seaplane that would be able to provide fighter cover during the early parts of amphibious landings. With the success of the A6M2 Zeke fresh in their minds, they instructed the Nakajima company to develop a seaplane version of the A6M2 Zeke that they were building for Mitsubishi.

Using the airframe of the A6M2, they replaced the landing gear with a large central float attached to the underside of the fuselage by a forward-sloping single strut in the front and a V-shaped strut at the rear. Beneath the wings, towards the tips, two stabilising cantilever floats were fitted. On the land version a drop fuel tank could be fitted beneath the fuselage, but because the struts for the central float were in the way it was decided to use the float itself as an additional fuel tank. On 7 December 1941, the Mitsubishi A6M2-N Navy Type 2 Floatplane Fighter Model 11 made its maiden flight, and within weeks production started.

Code-named 'Rufe' by the Allies, it proved to be a fast, manoeuvrable and reliable fighter aircraft and saw action in the Solomon Islands and in the Aleutian Campaign. However it turned out to be no match for the Allied fighters that it came up against, especially the Lockheed P-38 Lightnings during the Aleutian Campaign. Towards the end of the war, the Nakajima-built A6M2-N was used as a training aircraft for pilots who were to fly the Kawanishi N1K1 Kyofu. Between December 1941 and September 1943, a total of 327 Nakajima-built A6M2-Ns were built.

VERY LONG-RANGE FLYING BOATS

With the success of the new aircraft, the Navy decided to expand the size of its aircraft and work began on the development of two types of larger flying boat. One of Kawanishi's top designers, Yoshio Hashiguchi, was sent to Short Brothers in Britain to help in the design. The first one of the two types to be built was based on the Short Singapore and Calcutta models. Powered by three 825-hp Rolls-Royce Buzzard engines that were strut-mounted between the wings, the flying boat was tested in Britain, then dismantled and shipped to Japan by sea. It was evaluated by the Navy and accepted, and permission was given for the aircraft to be built by Kawanishi under licence.

Four of the flying boats were built and given the designation Kawanishi Type 90-2 (H3K1). There were a number of minor differences between all four of the aircraft, but externally they looked almost identical. They were of a unique design for the time inasmuch as the hull was all-metal and the lower part was of stainless steel. The hull also incorporated a tail-gun turret: the first time one had ever been installed in a Japanese aircraft.

The H3K1 was the largest flying boat in the Pacific and a demonstration of its capability was carried out in the summer of 1932, when Lieutenant Sukemitsu Itoh and his crew made a long-range flight between Tokyo Bay and Saipan, a distance of 1,300 nautical miles. This flight established the Kawanishi company as the number one manufacturer of flying boats in Japan.

There was a second design produced and one prototype was manufactured by the Hiro Arsenal, but was not accepted by the Navy although it was used as a flying test bed for various engines. With the new land-based trainer now operational, it was decided to replace the existing Type 13 Sea Trainer. The Yokosuka Arsenal was tasked with the new design and Lt-Com. Jiro Saha and Engineer Tamefumi Suzuki led the team. The aircraft produced was radical inasmuch as the fuselage was of a welded steel tube construction, and it was the first time this had ever been tried before in Japan. The wings and tail were made of wood and covered with fabric, while the floats were made of metal. Powered by an 130-hp inverted 7-cylinder, inline, air-cooled radial engine fitted with a two-bladed wooden propeller made by the Hatakaze company, the Navy Type 90 (K4Y1), as it was designated, had a top speed of 101 mph. Two prototypes were built and sent to the Navy for test and evaluation. Both were accepted, but with the proviso that the engine was changed to a 130-hp Gasuden Jimpu.

A contract was then put out to the Watanabe company and the Nippi company to build the aircraft. A total of 211 Navy Type 90 (K4Y1) were built and delivered to the Navy. Between 1932 and 1939, 156 were built by the Watanabe company, and between 1939 and 1940 fifty-three were built by the Nippi company.

During tests of the second prototype, a number of Army officers were present, and they were very impressed with the performance of the Ka-14; so much so that they ordered an almost identical aircraft under the designation Ki-18. The prototype was tested at Tachikawa by Army test pilots and was found to be faster than any of the existing Army fighters, but less manoeuvrable. One of these was the Ki-10, so the Koku Hombu returned the aircraft to Mitsubishi, asking for a prototype to be developed that had all the characteristics of the Ki-18 but with the manoeuvrability of the Ki-10. This resulted in the Ki-33, which will be dealt with later.

By 1937 the Navy had formed the Thirteenth Air Corps, with a total of thirty-nine land-based squadrons and 563 aircraft. At sea they had over 330 aircraft operating from aircraft and seaplane carriers. The addition of two more aircraft carriers, *Soryu* and *Hiryu*, and three seaplane carriers, *Chitose*, *Chiyoda* and *Mizubo*, brought the number of aircraft carriers to six and the number of seaplane carriers to five. It was around this time that experiments were being carried out on the use of aircraft on submarines, primarily for reconnaissance, but later for bombing.

The IJAAF had learned lessons from the war in Manchuria, one of which was that they had no long-range bombers of any consequence. So in 1933 they gave contracts to Mitsubishi and Kawasaki for three new bombers. Mitsubishi was tasked with the design and development of one heavy bomber and one light bomber, while Kawasaki was tasked with one light bomber.

Mitsubishi had imported a Junkers K37 in 1931, and had used it during the war in Manchuria. The company decided to base its new design for the heavy bomber on this aircraft as the specifications laid down by the Army required the bomber to be a twin-engined low-winged monoplane, capable of carrying a bomb load between 1,000 kg and 1,500 kg. The chief designer at the time was Nobishiro Nakata, and he produced a design that was, in essence, an enlarged version of the Junkers K37. Two 800-hp Rolls-Royce Buzzard engines powered the prototype. It had been intended to fit two Type 93 700-hp Mitsubishi water-cooled engines, but they had had been delayed in their development. After exhaustive tests the bomber was given the designation Mitsubishi Army Type 93 (Ki-1) Heavy Bomber and was put into production, despite the problems with the engines. The main problem was that level flight could not be maintained if one of the engines failed.

In comparison to the Junkers K37, the Type 93 Heavy Bomber had its engines mounted lower on the wings, an extended undercarriage and the corrugated panels on the wings replaced with a smooth metal skin, although the corrugated skin was retained on the fuselage. It had two pilots, a bombardier, and three 7.7 mm flexible machine gun positions, one in the nose, one dorsal and one ventral. In the following three years, 118 of the aircraft were built.

TWO SUCCESSFUL BOMBERS

In 1933, Admiral Isoroku Yamamoto, who at the time was the head of the Technical Division of the Naval Bureau of Aeronautics, put forward a proposal for the design and development of a land-based, long-range bomber. He had the foresight to see that the vastness of the Pacific Ocean could be turned to Japan's advantage if it had such an aircraft. At the time they only had a few aircraft carriers, and land bases were spread over a wide area.

Mitsubishi was approached and provided with the requirements on a non-competitive basis. The aircraft was to be a prototype that could be assessed, by making various modifications, as a future long-range bomber. The designers were not hampered by specific military requirements, which in essence gave them a blank sheet on which to work.

In April 1934 the prototype Ka-9, as it was called, emerged from the Mitsubishi factory. It was a twin-engined aircraft with the wings mounted mid-fuselage, and with a Junkers-style tail consisting of twin fins and rudders. The first flight went without a hitch and demonstrated a large bomber with good handling characteristics and exceptional manoeuvrability. The aircraft was then handed over for further testing and evaluation, and during one test the Navy carried out a 3,760-mile (3,265 nautical miles) test flight.

The Navy, understandably, was delighted with the results and issued another proposal, this time with a full specification calling for an attack bomber capable of carrying a bomb load of 1,764 lb (800 kg) and a defensive armament of three 7.7 mm machine guns. As with the initial proposal, this was issued on a non-competitive basis as the need for a long-range bomber suddenly became more urgent.

The new model, designated the Mitsubishi Ka-15, had a larger fuselage with space for three retractable turrets, two dorsal and one ventral, in which the 7.7 mm machine guns would be housed. The aircraft was of an all-metal construction with fabric-covered control surfaces. The tail surfaces were enlarged to improve stability during bombing runs. Provision was also made to carry a 1,764 lb (800 kg) torpedo beneath the fuselage. The first of the prototypes rolled out of the factory in July 1935, and the first test flight was made. The Navy was delighted with the aircraft and realised that they now had a long-range bomber the equal of any in the west.

Over the next year another twenty prototypes were built, all with minor modifications and a variety of engines. Then, in June 1936, the Navy ordered the aircraft into production and gave the designation of the Navy Type 96 Attack Bomber Model 11 (G3M1). Powered by two 680-hp Mitsubishi Kinsei 2 14-cylinder radial engines, the first twenty were known as the 'solid nose' version because of the unglazed nose section. The next twenty were powered with the Mitsubishi Kinsei 3 engines and had a glazed nose section. A number of different models appeared over the next few months and one, the G3M2, had larger fuel tanks installed and two 1,075-hp Mitsubishi Kinsei engines fitted. This increased the power and the range of the bomber considerably.

Then, a week after the second Sino-Japanese conflict started, the G3M2 made its appearance, leaving the base at Taipei (Formosa) to bomb targets in the Hangchow and Kwangteh areas on the mainland of China. They flew 1,250 miles over the ocean and, despite bad weather conditions, carried out the first of a number of bombing raids. The following day a second raid, G3M2s operating from their base on Kyushu, attacked the Chinese mainland. There was a downside to these raids, inasmuch as they carried out these missions deep into Chinese territory without a fighter escort, and as such became vulnerable to attack by enemy fighters. Their defensive armament was wholly inadequate and they suffered heavy losses.

By the summer of 1940, the Mitsubishi company had built 343 G3M2s and the Nakajima company was beginning to produce the aircraft under contract from the Navy. Because there were very few high-performance aircraft around at the time, twenty-four G3M2s were converted into Mitsubishi twin-engined transports for Japan Air Lines (Nippon Koku KK). These civil versions made a number of record-breaking flights from Tokyo to Rome and Tehran. There was even a round-the-world flight covering a distance of 32,850 miles in 194 hours, a remarkable achievement for the time.

A number of variants appeared at the time, most with increased offensive armament. A large 'turtle-back' turret replaced both the ventral and rear dorsal retractable turrets, and provision was made for another 7.7 mm machine gun to be mounted in the cockpit in such a way that it could be fired from either side.

Other equipment changes incorporated the fitting of the Sperry automatic pilot, which was manufactured under licence, and a radio direction finder unit, both of which were later adopted as standard in the Model 22.

When the Pacific War started, the Japanese Navy had 204 G3M2s operating in front-line units, and a further fifty-four in second-line units. The aircraft was in the forefront of the attacks against the Americans in the assaults on the Philippines, the Marianas and Wake Island. Their first major success, however, came when sixty G3M2s, accompanied by twenty-six G4M1s, attacked two British battleships, HMS *Prince of Wales* and HMS *Repulse*, off the coast of Malaya and sank them both. This was a serious setback to the Royal Navy and a major victory for the Japanese.

As the war progressed the G3M was replaced by the G4M1. Although still being used operationally, they were gradually relegated to other roles such as glider tugs and bomber trainers. A small number were fitted with search radar and additional fuel tanks and used as maritime reconnaissance aircraft.

Between 1935 and 1941, a total of 1,084 G3Ms were built by the Mitsubishi and Nakajima companies.

One of the best known of all the Japanese bombers was the Mitsubishi G4M, known to the Allies as 'Betty'. Developed in September 1937 in response to a request by the Navy for a twin-engined, land-based attack bomber, the G4M was intended as a replacement for the Navy Type 96 Attack Bomber, the G3M1. The G4M's lack of armour for its crew and fuel tanks also earned itself the name of the 'Flying Lighter' because of its tendency to burst into flames when raked by machine gun fire. The G4M was the brainchild of the designer Kiro Honjo, who was asked to produce a bomber that had both range and speed using

2,000-hp engines. The only way to achieve this was to make the aircraft a light as possible without sacrificing its armament and bomb-carrying ability.

The fat, semi-monocoque, cigar-shaped fuselage was designed to allow the crew to move freely within it, and a bomb bay was fitted beneath the wing centre section. The wings were mounted mid-fuselage and contained two unprotected fuel tanks containing 1,078 gallons (4,900 litres) of fuel. The armament consisted of one 7.7 mm machine gun mounted in a nose-cone that rotated mechanically through 360 degrees of the aircraft's axis; one flexibly mounted 7.7 mm machine gun in a dorsal blister behind the cockpit; one 7.7 mm machine gun in blisters either side of the fuselage, just behind the trailing edge of the wings, and one 20 mm cannon in a tail turret.

After initial flight tests in 1939, the twin engines were replaced with 1,530-hp Mitsubishi Kasei 11 14-cylinder, air-cooled radials. The aircraft performed better than expected and easily exceeded the requirements laid down by the Navy. The G4M had a top speed of 276 mph (240 kt) and a range of 3,453 statute miles (3,000 nautical miles). Mitsubishi waited for the Navy to place an order for the aircraft, but instead was asked for a heavy escort fighter version, the G6M1, because there were no escort fighters capable of escorting their Navy Type 96 Attack Bombers. To accommodate the Navy, Mitsubishi faired over the bomb bay and placed a ventral gondola in its place containing two 20 mm cannons, one rearward-firing and one forward-firing, and the two machine guns in the fuselage side blisters were replaced with 20 mm cannons. The dorsal turret was removed, although the nose and tail machine guns were retained.

In addition to the heavy armament being installed, a large amount of ammunition was carried in the fuselage, which created extra weight. This, of course, required additional fuel so that the engines could maintain the take-off weight of 20,944 lb (9,500 kg). With these modifications in place, the Mitsubishi G6M1, as it was called, was assigned its new role. However, it was quickly established that because of the additional weight, the combat speed of the aircraft was slower than the aircraft it was supposed to be protecting.

Realising their mistake, the Navy took the aircraft away from that role and further modified it as a crew trainer (G6M1-K), and also a Navy Type 1 Paratroop Transport (G6M-L2). With this done, authorisation came through in 1940 to finally start production of the G4M1. The first sixteen of the aircraft appeared in April 1941 and were immediately placed with the First Kokutai in China.

Early in 1938, the JAAF was looking for a replacement for its Army Type 97 Heavy Bomber (Ki-21) and approached the Mitsubishi company with a proposal for a long-range heavy bomber capable of operating without escort fighters and able to look after itself in the event of it being attacked. Nakajima gave the project to one of the senior engineers and work began in designing an aircraft that would match the specifications laid down by the Army.

This aircraft had to be capable of operating without a fighter escort, relying on speed and heavy armament. It had to have a maximum speed of 310 mph (500 km/h), a range of 1,864 miles (3,000 km), a bomb load of 2,205 lb (1,000 kg) and a range of armament including a flexible 20 mm cannon in a dorsal turret, two 7.7 mm machine guns mounted in the nose, and a 7.7 mm machine gun in a tail turret.

Using the knowledge gained from building the Ki-21, the engineers produced a twin-engined bomber with mid-mounted wings that contained six self-sealing fuel tanks, three on each side. In an effort to reduce drag, the engine nacelles were mounted well ahead of the flaps. The bombs were carried in a bomb bay that was the entire length of the wing centre-section.

Powered by two 950-hp Nakajima Ha-5 KAI air-cooled radial engines, the first Nakajima Ki-49-I Donryu prototype was subjected to trials in August 1939 and was well received by the test pilots for its handling and manoeuvrability. The next two prototypes were fitted with 1,250-hp Nakajima Ha-41 engines, and these were followed by a further seven service production models, which were submitted to the Army for test and evaluation. Minor modifications were made as a result of these tests and the aircraft went into production.

The bomber was given the code name 'Helen' by the Allies, and first saw action during the Pacific War over Australia's Northern Territories. The Japanese pilots, who had been flying the Ki-21 in China, found that the Ki-49 was more difficult to fly under battle conditions than its predecessor. It may have been faster, but it was less manoeuvrable and carried a smaller bomb load, although there were favourable comments regarding the self-sealing fuel tanks and the defensive armour in the cockpit area.

At the beginning of 1942 it was decided to install a pair of 1,450-hp Nakajima Ha-109 air-cooled radial engines. The engines were almost the same size of the Ha-41 engines, the only difference being that the oil cooler that was mounted on the front of the Ha-41 was moved to beneath the Ha-109. A number of other minor modifications were made, including thicker armour plating for the crew. The aircraft was designated the Ki-49-II, and two pre-production models were produced for evaluation. Production was started almost immediately and the first of the Model IIs came off the production line in August 1942.

The aircraft still didn't measure up to what was required, and the armament carried by the Ki-49-I and II was almost ineffective against the Allied fighters. The bombers were sent to operate in the New Guinea/China/Manchuria theatres of the war, where they did not come up against the strong opposition of Allied fighters. Pilots complained about the lack of speed at lower and medium altitudes and considered the older Ki-21-IIs to be far superior.

In an effort to improve the aircraft the more powerful 2,800-hp Nakajima Ha-117 air-cooled, radial engines were fitted, but they suffered from endless teething problems and only six Model IIIs with theses engines were ever built. A number of other uses were found for Ki-49-IIs, such as night fighters, anti-submarine roles and even as a troop transport, but none were successful and, as with most of the Japanese aircraft towards the end of the war, they became used for kamikaze missions. Its one claim to fame was that it was the first Japanese bomber to be fitted with a tail turret.

A JAPANESE AUTOGIRO

One of the strangest aircraft produced by the Japanese was the Kayaba Ka-1. This was an autogiro and was the first and only time that an autogiro was used operationally, armed with two 132 lb (60 kg) depth charges. Towards the end of the 1930s the Japanese Army had showed an interest in using the autogiro as an artillery spotter, and had purchased a Kellet KD-1A from America. During low-level flight trials the aircraft crashed and was destroyed. Not to be deterred, the Army sent the wreckage to the Kayaba Industrial Company, which was known to be experimenting in the development of the autogiro. They were instructed by the Army to build a similar machine using what information they could from the wreckage.

The company developed a two-seat autogiro, powered by a 240-hp Argus As 8-cylinder air-cooled engine that had a three-bladed rotor mounted atop a pylon on the fuselage in front of the pilot. The first test flight went remarkably well, that aircraft demonstrating the ability to hover almost motionless and execute a 360-degree turn at the same time. This was achieved by running the engine at full power and holding the nose of the aircraft at an upward 15-degree angle.

The aircraft was ordered into production. As they came off the production line, they were assigned to artillery units. As the war in the Pacific progressed and Japanese shipping losses started to mount, it was decided to use the Kayaba Ka-1 for anti-submarine patrols. It was soon discovered that the engine was not powerful enough to sustain two crewmembers and two depth charges, so they were converted into single-seat aircraft and powered by a 240-hp Jacobs L-4Ma-7, 7-cylinder, air-cooled radial engine.

These anti-submarine aircraft operated from a converted merchant ship, the *Akitsu Maru*, patrolling the coastal areas and concentrating on Korean and Tsugara channels. A total of 240 Kayaba Ka-1 autogiros were built between 1939 and 1942.

Just before the outbreak of the Second World War, the Bücker Flugzeubau company demonstrated their primary two-seat trainers, the Bü 131B Jungman and the Bü 133C Jungmeister, to the Japanese. The Japanese military was so impressed that it purchased one of each and subjected them to an intense testing programme. Everyone who flew the aircraft was delighted with them, and the ground maintenance crews passed favourable comments on their simplicity of construction and the ease of maintenance.

In 1939 the Navy purchased twenty of the aircraft and tested them under training conditions, subjecting them to the rigours of being flown by novice pilots. The aircraft passed the tests with flying colours, so the Navy instructed Watanabe to design a similar aircraft based on the design of the Bü 131, but not to copy it. The result was very disappointing, so much so that the company sent one of their top engineers to Germany to negotiate the manufacturing rights. In 1942, with the rights negotiated, the aircraft was put into production and given the designation Navy Type 2 Primary Trainer Model 11 (K9W1). The aircraft became the Navy's standard primary trainer. The Navy had a total number of 339 of the aircraft built between 1942 and 1944.

The success of the aircraft attracted the attention, and in 1943 the Army requested the Nippon Kokysai Koku Company manufacture a version of the aircraft. The Army version was of an all-wood construction with fabric-covered wings. Given the designation Ki-86 by the Army, a total of 1,037 of the aircraft were built between the beginning of 1943 and 1945.

SPECIFICATIONS

Aichi D1A

Wing Span:	37 ft 3½ in. (11.37 m) (D1A1)
	37 ft 4¾ in. (11.4 m) (D1A2)
Length:	30 ft 10¼ in. (9.4 m) (D1A1)
	30 ft 6½ in. (9.3 m) (D1A2)
Height:	1 ft 2¾ in. (3.45 m) (D1A1)
	11 ft 2¼ in. (3.41 m) (D1A2)
Weight Empty:	3,086 lb (1,400 kg) (D1A1)
	3,342 lb (1,516 kg) (D1A2)
Weight Loaded:	5,291 lb (2,400 kg) (D1A1)
	5,512 lb (2,610 kg) D1A2)
Max. Speed:	174 mph (151 knt) (D1A1)
	192 mph (167 knt) (D1A2)
Ceiling:	22,965 ft (7,000 m)
Range:	656 miles (570 naut/m)
Endurance:	Not known
Engine:	One 580-hp Nakajima Kotobuki 2, 9-cylinder, air-cooled radial (D1A1)
	One 715-hp Nakajima Kotobuki 3, 9-cylinder, air-cooled radial (D1A2)
Armament:	Four flexible 7.7 mm machine guns, one 551 lb (250 kg) bomb under fuselage, two 66 lb (30 kg) bombs beneath the wings

Nakajima Ki-49 Donryu (Storm Dragon) (Helen)

Wing Span:	67 ft 0¼ in. (20.4 m)
Length:	55 ft 1¼ in. (16.8 m)
Height:	13 ft 11¾ in. (4.25 m)
Weight Empty:	13,382 lb (6,070 kg)
Weight Loaded:	22,377 lb (10,150 kg)
Max. Speed:	306 mph (492 knts)
Ceiling:	30,510 ft (9,300 m)
Range:	1,833 miles (2,950 km)
Endurance:	Not known
Engine:	Two 950-hp Nakajima Ha-5, 14-cylinder, air-cooled radials
Armament:	One flexible 20 mm cannon in dorsal turret, one flexible 12.7 mm machine gun in the nose, ventral, port and starboard beams and tail positions

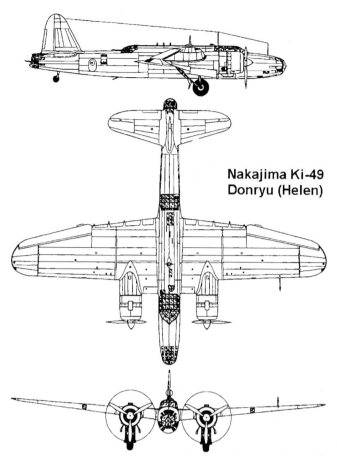

**Nakajima Ki-49
Donryu (Helen)**

Nakajima Ki-49 Donryu
('Storm Dragon')
– Helen.

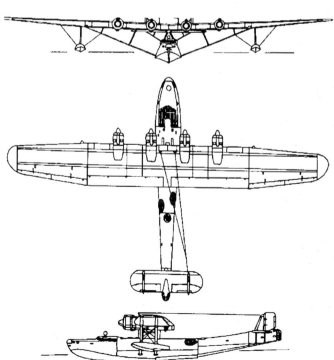

Kawanishi H6K Flying
Boat.

KAWANISHI H6K

Wing Span:	131 ft 2¾ in. (40.0 m)
Length:	84 ft 0¾ in. (25.6 m)
Height:	20 ft 6¾ in. (6.25 m)
Weight Empty:	22,796 lb (10,340 kg)
Weight Loaded:	35,274lb. (16,000 kg)
Max. Speed:	206 mph (179 knts)
Ceiling:	24,935 ft (7,600 m)
Range:	2,567 stat. miles (2,230 naut. miles)
Endurance:	Not known
Engine:	Four 840-hp Nakajima Hikari 2, 9-cylinder, air-cooled radials
Armament:	One flexible 7.7 mm machine gun in dorsal turret, one flexible 7.7 mm machine gun in bow position and one 7.7 mm machine gun in tail turret; two 1,764 lb (800 kg) torpedoes or one 2,250 lb (1,000 kg) bomb
Crew:	Eight, with provision for ten passengers

KAWANISHI H8K

Wing Span:	124 ft 8¼ in. (38.0 m)
Length:	92 ft 3¾ in. (28.2 m)
Height:	30 ft 0¼ in. (9.15 m)
Weight Empty:	34,176 lb (15,502 kg)
Weight Loaded:	54,013 lb (24,500 kg)
Max. Speed:	269 mph (234 knts)
Ceiling:	25,035 ft (7,630 m)
Range:	4,475 st. miles (3,888 naut. miles)
Endurance:	Not known
Engine:	Four 1,530-hp Mitsubishi Kasei 11, 14-cylinder, air-cooled radials.
Armament:	One 20 mm cannon in dorsal, bow and tail turrets, two 7.7 mm machine guns in blisters either side of the fuselage and cockpit hatches; two 1,764 lb (800 kg) torpedos, or eight 551 lb (250 kg) bombs
Crew:	Ten

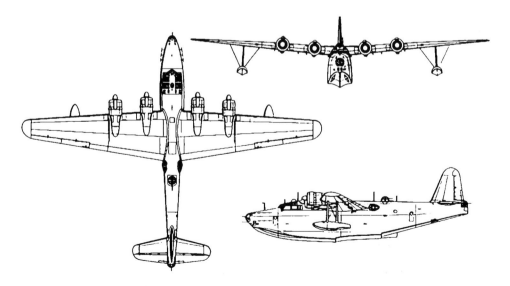

Kawanishi H8K Flying Boat

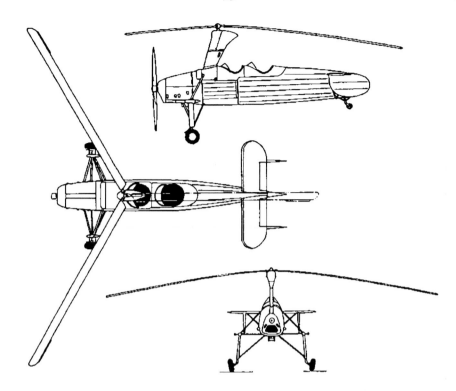

Kayaba Ka-1 Autogiro.

KAYABA KA-1

Length: 30 ft 2¼ in. (9.2 m)
Height: 10 ft (3.05 m)
Rotor diameter: 40 ft 0¼ in. (12.2 m)
Weight Empty: 1,709 lb (775 kg)
Weight Loaded: 2,579 lb (1,170 kg)
Max. Speed: 102mph (165 km/h)
Ceiling: 11,485 ft (3,500 m)
Endurance: Not known
Range: 174 miles (280 km)
Engine: 240-hp Argus As 10c, 8-cylinder, air-cooled
Armament: Two 132 lb (60 kg) depth charges

NAKAJIMA NAVY TYPE 95 CARRIER FIGHTER (A4N1)

Wing Span: 32 ft 9½ in. (10 m)
Length: 21 ft 9¼ in. (6.64 m)
Height: 10 ft 1 in. (3.07 m)
Weight Empty: 2,813 lb (1,276 kg)
Weight Loaded: 3,880 lb (1,760 kg)
Max. Speed: 218 mph (190 kts)
Ceiling: 25,393 ft (7,740 m)
Endurance: 3½ hours
Engine: 80-93-hp Le Rhône, 9-cylinder, air-cooled rotary
Armament: One forward-firing 7.7 mm machine gun; two 66 lb (30 kg) or one 132 lb
                   (60 kg) bombs

NAKAJIMA ARMY TYPE 94 RECONNAISSANCE AIRCRAFT (KI-4)

Upper Wing Span:	39 ft 4½ in. (12 m)
Lower Wing Span:	26 ft 7¼ in. (8.11 m)
Length:	25 ft 4¼ in. (7.73 m)
Height:	11 ft 5¾ in. (3.50 m)
Weight Empty:	3,668 lb (1,664 kg)
Weight Loaded:	5,454 lb (2,474 kg)
Max. Speed:	176 mph (153 kts)
Ceiling:	26,246 ft (8,000 m)
Endurance:	Not known
Engine:	600-640-hp Nakajima Ha-8, 9-cylinder, air-cooled rotary
Armament:	Two fixed forward-firing 7.7 mm machine guns, one flexible dorsal-mounted 7.7 mm machine gun

WATANABE E9W1

Wing Span:	33 ft 9 in. (9.9 m)
Length:	25 ft ¾ in. (7.64 m)
Height:	10 ft 9½ in. (3.29 m)
Weight Empty:	1, 867 lb (847 kg)
Weight Loaded:	2,667 lb (1,210 kg)
Max. Speed:	145 mph (126 kts)
Ceiling:	22,112 ft (6,740 m)
Endurance:	5 hours.
Range:	359 naut. miles (454 miles)
Engine:	One 340-hp Gasuden Tempu 11, 12–cylinder, air-cooled, radial
Armament:	One dorsal 7.7 mm machine gun
Crew:	Two

NAKAJIMA ARMY TYPE 95-2 TRAINER (KI-6)

Wing Span:	50 ft 7¾ in. (15.5 m)
Length:	36 ft 4½ in. (11.09 m)
Height:	9 ft 3 in. (2.81 m)
Weight Empty:	3,615 lb (1,640 kg)
Weight Loaded:	5,952 lb (2,700 kg)
Max. Speed:	153 mph (133 kts)
Ceiling:	19,684 ft (6,000 m)
Endurance:	5½ hours
Range:	651 st. miles (566 naut. mls)
Engine:	One 450-580-hp Nakajima Jupiter VII, 9-cylinder, air-cooled radial
Armament:	One dorsal flexible 7.7 mm machine gun

AICHI D3A

Wing Span:	47 ft 1¾ in. (14.3 m)
Length:	33 ft 5½ in. (10.19 m)
Height:	12 ft 7¾ in. (3.84 m)
Weight Empty:	5,309 lb (2,408 kg)
Weight Loaded:	8,047 lb (3,650 kg)
Max. Speed:	240 mph (209 kts)
Ceiling:	30,050 ft (9,030 m)
Endurance:	Not known

Range:	915 st. miles (795 naut. mls)
Engine:	One 800-hp Nakajima Kikari 1, 9-cylinder, air-cooled radial
Armament:	Two fixed forward-firing 7.7 mm machine guns mounted in the engine cowling and one flexible rearward-firing 7.7 mm machine gun
Crew:	Two in tandem

NAKAJIMA NAVY TYPE 90-2-2 RECONNAISSANCE SEAPLANE (E4N2)

Wing Span:	36 ft (10.97 m)
Length:	29 ft 1¼ in. (8.87 m)
Height:	13 ft (3.96 m)
Weight Empty:	2,760 lb (1,252 kg)
Weight Loaded:	3,968 lb (1,800 kg)
Max. Speed:	144 mph (125 kts)
Ceiling:	18,832 ft (5,740 m)
Endurance:	5 hours
Range:	633 sq. miles (550 naut. mls)
Engine:	One 460-580-hp Nakajima Kotobuki 2, 9-cylinder, air-cooled, radial
Armament:	One dorsal flexible 7.7 mm machine gun, two 66 lb (30 kg) bombs

KAWASAKI KI-45 (TORYU)

Wing Span:	47 ft 6¾ in. (14.5 m)
Length:	33 ft 8 in. (10.26 m)
Height:	11 ft 8½ in. (3.57 m)
Weight Empty:	5,512 lb (2,500 kg)
Weight Loaded:	8,267 lb (3,750 kg)
Max. Speed:	298 mph (480 km/h)
Ceiling:	35,200 ft (2,260 m)
Endurance:	Not known
Range:	1,087 sq. miles (1,750 km)
Engine:	Two 730-hp Nakajima Ha 2b, 9-cylinder, air-cooled, radials
Armament:	Two nose-mounted 12.7 mm machine guns, two dorsal flexible 20 mm machine guns, one rear-firing flexible 7.7 mm machine gun and one starboard side ventral 20 mm machine gun; two 551 lb (250 kg) bombs

NAKAJIMA KI-27 FIGHTER

Wing Span:	37 ft 1¼ in. (11.3 m)
Length:	24 ft 8½ in. (7.53 m)
Height:	10 ft 7½ in. (3.25 m)
Weight Empty:	2,447 lb (1,110 kg)
Weight Loaded:	3,946 lb (1,790 kg)
Max. Speed:	292 mph (470 km/h)
Ceiling:	16,405 ft (5,000 m)
Endurance:	Not known
Range:	1,060 miles (1,710 km)
Engine:	Two 780-hp Nakajima Ha 1b, 9-cylinder, air-cooled, radial
Armament:	Two synchronised 12.7 mm machine guns; four 55 lb (25 kg) bombs

NAKAJIMA KI-34 TRANSPORT

Wing Span:	65 ft ¼ in. (19.9 m)
Length:	50 ft 2½ in. (15.3 m)

Height: 13 ft 7½ in. (4.15 m)
Weight Empty: 7,716 lb (3,500 kg)
Weight Loaded: 11,574 lb (5,250 kg)
Max. Speed: 224 mph (360 km/h)
Ceiling: 22,965 ft (7,000 m)
Endurance: Not known
Range: 745 miles (1,200 km)
Engine: Two 710-hp Nakajima Kotobuki, 9-cylinder, air-cooled, radials
Armament: None
Crew: Three, plus eight passengers

NAKAJIMA KI-43 FIGHTER/BOMBER (OSCAR)

Wing Span: 37 ft 6¼ in. (11.4 m) (Ia)
                     35 ft 6¾ in. (10.84 m) (IIb & IIIa)
Length: 28 ft 11¾ in. (8.83 m) (Ia)
                     29 ft 3¼ in. (8.92 m) (IIb & IIIa)
Height: 10 ft 8¾ in. (3.27 m) (Ia, IIb & IIIa)
Weight Empty: 3,483 lb (1,580 kg) (Ia)
                     4,211 lb (1,910 kg) (IIb)
                     4,233 lb (1,920 kg) (IIIa)
Weight Loaded: 4,515 lb (2,048 kg) (Ia)
                     5,710 lb (2,590 kg) (IIb)
                     5,644 lb (3,060 kg) (IIIa)
Max. Speed: 308 mph (495 km/h) (Ia)
                     329 mph (530 km/h) (IIb)
                     358 mph (576 km/h) (IIIa)
Ceiling: 38,500 ft. (11,750 m) (Ia)
                     36,750 ft (11,200 m) (IIb)
                     37,400 ft (11,400 m) (IIIa)
Endurance: Not known
Range: 1,990 miles (3,200 km) (Ia & IIb & IIIa)
Engine: One 710-hp Nakajima Ha-115, 14-cylinder, air-cooled, radial (Ia, IIb & IIIa)
Armament: Two 7.7 mm machine guns (Ia)
                     One 7.7 mm machine gun and one 12.7 mm machine gun (IIb)
                     Two 12.7 mm machine guns (IIIa)
                     Two 33 lb (15 kg) bombs (Ia)
                     Two 66 lb (30 kg) bombs or one 551 lb (250 kg) bomb
Crew: One

TACHIKAWA KI-36 & KI-55

Wing Span: 38 ft 8½ in. (11.8 m) (Ki-36&55)
Length: 26 ft 2¾ in. (8.0 m) (Ki-36&55)
Height: 11 ft 11½ in. (3.64 m) (Ki-36&55)
Weight Empty: 2,749 lb (1,247 kg) (Ki-36)
                     2,848 lb (1,292 kg) (Ki-55)
Weight Loaded: 3,660 lb (1,660 kg) (Ki-36)
                     3,749 lb (1,721 kg) (Ki-55)
Max. Speed: 217 mph (349 km/h) (Ki-36&55)
Ceiling: 26,740 ft (8,150 m) (Ki-36)
                     26,900 ft (8,200 m) (Ki-55)
Endurance: Not known
Range: 767 miles (1,235 km) (Ki-36)

659 miles (1,060 km) (Ki-55)

Engine: One 450-hp Hitachi Ha 13a, 9-cylinder, air-cooled, radial
Armament: One forward-firing 7.7 mm machine gun mounted in the engine cowling
               and one flexible rear-firing 7.7 mm machine guns; two 551 lb (250 kg)
               bombs; one 1,102 lb (500 kg) bomb on kamikaze sortie

NAKAJIMA B5N NAVY TYPE 97 CARRIER ATTACK BOMBER (KATE)

Wing Span: 36 ft 0¼ in. (10.98 m)
Length: 28 ft 10¾ in. (8.8 m)
Height: 12 ft 7¼ in. (3.8 m)
Weight Empty: 2,910 lb (1,320 kg)
Weight Loaded: 4,189 lb (1,900 kg)
Max. Speed: 186 mph (162 knts)
Ceiling: 23,850 ft (7,270 m)
Endurance: Not known
Range: 558 miles (484 naut. mls)
Engine: One 630-hp Nakajima Kotobuki 2, 9-cylinder, air-cooled, radial
Armament: One fixed forward-firing 7.7 mm machine gun and one flexible rearward-
                firing 7.7 mm machine gun
Crew: Two

YOKOSUKA B4Y NAVY TYPE 96 CARRIER ATTACK BOMBER (JEAN)

Wing Span: 49 ft 2¼ in. (15 m)
Length: 33 ft 3½ in. (10.15 m)
Height: 14 ft 3¾ in. (4.36 m)
Weight Empty: 4,409 lb (2,000 kg)
Weight Loaded: 7,937 lb (3,600 kg)
Max. Speed: 173 mph (150 knts)
Ceiling: 19,685 ft (6,000 m)
Endurance: Not known
Range: 978 miles (850 naut. mls.)
Engine: One 840-hp Nakajima Hikari 2, 12-cylinder, air-cooled, radial
Armament: One flexible rearward-firing 7.7 mm machine gun
Crew: Three

KYUSHU K9W1 & KOKUSAI KI-86 (CYPRESS)

Wing Span: 24 ft ¾ in. (7.34 m)
Length: 21 ft 8½ in. (6.61 m)
Height: 8 ft 7¾ in. (2.63 m)
Weight Empty: 902 lb (409 kg)
Weight Loaded: 1,409 lb (639 kg)
Max. Speed: 112 mph (180 knts)
Ceiling: 12,730 ft (3,880 m)
Endurance: Not known
Range: 373 miles (600 km)
Engine: One 110-hp Hitachi Ha-47, 14-cylinder, air-cooled, in-line
Armament: None
Crew: Two

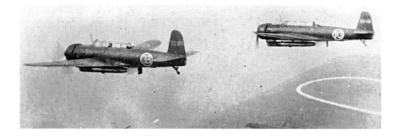

Nakajima B6N2s
in formation.

Nakajima B6N2
about to take off.

Aichi H9A-1
amphibian.

Aichi H9A Type 2
taxiing in towards
slipway.

Warrant Flying Officer Nobuo Fujita.

Petty Officer Shoji Okuda.

Yokosuka E14Y.

Yokosuka E14Y1 about to be launched from a submarine.

Nakajima C6N-1

Nakajima Ki84.

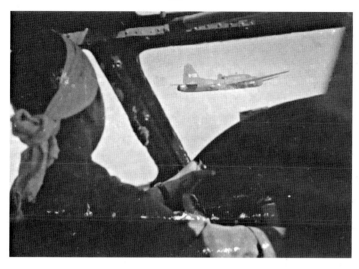

Pilot of a Mitsubishi Ki-109 keeping in formation during a bombing mission.

Mitsubishi Ki-67-I.

Yokosuka Ohka flying bomb.

Kawanishi N1K2-J.

Tachikawa Ki-36 being prepared for a mission.

Tachikawa Ki-36 about to take off.

Nakajima Ki-43-II about to take off.

Nakajima Ki-43-42 in formation.

Nakajima Ki-43-45 with close-up of pilot in cockpit.

Kawasaki Ki61.

Aichi A6M2.

Hanger on board the I-400 submarine that housed two Aichi A6M2 aircraft.

Mitsubishi J8M rocket fighter.

The aircraft-carrying submarine I-26.

Lt Susumu standing beside his Mitsubishi J2M1 Raiden.

Mitsubishi Ki-109-2 with its crew.

Mitsubishi J2M1 Raiden.

Mitsubishi J2M3-21.

CHAPTER FOUR

1940s

At the end of 1939 the Koku Hombu realised that with the development of the bomber came the need for pilots who had been trained on multi-engined aircraft, and to this end they approached the Tachikawa company to design a twin-engined trainer. Not only was it to train pilots, but also the other members of a bombing crew, i.e. navigator, radio operator, bombardier and gunners.

Tachikawa's chief designer, Ryokichi Endo, produced a design for a low-wing monoplane with a retractable undercarriage, powered by two 510-hp Hitachi Ha-13a air-cooled radial engines. The prototype emerged just six months later in July 1940 and was tested by the Army. A number of minor modifications were made to correct the nose-heavy tendency the aircraft had demonstrated on landing.

At the beginning of 1941 the Tachikawa Ki-54a, as it was designated, was put into production as the Army Type Advanced Trainer Model A. The first models were primarily used for pilot training only, but the Model B, which was modified for crew training and had four gunnery stations, quickly followed this. The aircraft was assigned to all the military multi-engined training schools and radio schools, as well as a small number of civilian training schools.

So successful was the Ki-54 Model A and B that another version, the Model C, was created as a communication and light transport aircraft fitted with eight seats. Towards the end of the war another version appeared, the Model D. This was developed as an anti-submarine patrol aircraft and was fitted with facilities that enabled it to carry eight 130 lb (60 kg) depth charges.

Code-named 'Hickory' by the Allies, the Ki-54 was one of the most successful multi-engined trainers built by the Japanese, and a total of 1,368 of the aircraft were built between 1940 and 1945.

At the beginning of 1940, with the war clouds starting to gather, the JNAF decided the time was right to replace the ageing Nakajima B5N Navy Type 97 Carrier Attack Bomber. A tender was put out at the end of 1939 to the Nakajima company to design and build a replacement that called for a three-seat torpedo bomber that could operate from both the land and an aircraft carrier. It had to have a top speed of 288 mph (250 kts), and a range of 1,151 statute miles (1,000 nautical miles) carrying a bomb load of 1,764 lbs (800 kg).

With these specifications in mind, Kenichi Matsumura and his team set to work. Using the design of the B5N1 as a template, the Nakajima B6N Tenzan, as it was called, showed almost no difference or improvements to the B5N. The one thing that had to be re-designed were the wings; although they were the same size as the B5N, there were certain restraints placed on them because of the limited space aboard the aircraft carriers. Installing a 1,870-hp Nakajima NK7A Mamoru 11, 14-cylinder, air-cooled, radial engine provided additional power. The first of the prototypes appeared in May 1941, followed almost immediately by the second.

The first flight tests showed up a number of serious engineering defects, the main one being a stability problem that was resolved by moving the vertical tail surfaces 2 degrees 10 minutes to the left to counteract the torque produced by the large four-bladed propeller. With this problem resolved, that aircraft itself gave no more cause for concern, but teething problems with the engine slowed production dramatically. It wasn't until the end of 1942 that the Nakajima B6N2 Navy Carrier Attack Bomber, as it was designated, was ready for carrier trials.

The aircraft carriers selected for the trials were the *Ryuho* and the *Zuikaku* and the first take-offs went without a hitch, but the first landings showed a weakness in the arrester hook when they broke off after catching the arrester wire. Both prototypes were returned to the factory, where the hooks were strengthened. With a number of other modifications made, the remaining trials were carried out successfully and the aircraft was accepted by the JNAF. Amongst the other modifications were the strengthening of the undercarriage, a 2 degree downward angling of the torpedo racks beneath the starboard side of the fuselage, together with stabilising tail plates fitted to the torpedo itself to prevent it bouncing when dropped from a low altitude and the addition of a 7.7 mm machine gun that fired through a ventral tunnel.

The B6N2, however, did require RATO (Rocket Assisted Take-Off) when launching from a carrier with a full fuel, bomb or torpedo load. This weight problem and the high landing speed restricted the B6N2 to operating from the larger aircraft carriers. During the production programme another setback occurred when Nakajima were ordered to cease production of the Nakajima NK7A Mamoru 11 engine and concentrate on the manufacture of the Homare and Sakae engines, which were being more widely used. This meant that all subsequent models of the B6N2 Tenzan had to be re-engined with the 1,850-hp Mitsubishi MK4T Kasei 25, which was the engine the Navy initially wanted for the aircraft.

Two prototypes were produced, powered by the new engine and with a number of minor modifications being made. These included changing the retractable tail-wheel for a fixed one and replacing the 7.7 mm dorsal machine gun with a flexible 13 mm Type 2 machine gun.

As the war progressed, a number of other modifications were made to the B6N2, known as 'Jill' to the Allies, including the strengthening of the undercarriage, but by this time Japan had virtually no aircraft carriers and none that could take the Tenzan. The remaining models were relegated to carrying out kamikaze attacks during the bitter fighting on Okinawa. The Nakajima company built 1,268 B6N2 Tenzans between 1941 and 1945, and almost all saw action in the Pacific.

In December 1940, the increased use of flying boats prompted the Navy to issue a specification for the design of an advanced crew trainer. They wanted an aircraft that could train pilots and crews to the standard required to fly the high-performance Kawanishi H8K1 reconnaissance flying boat. This was unusual because up to that point all new aircraft had been designed and built specifically for combat roles, but this one had to be built for just training.

The Aichi company put forward a proposal for a twin-engined, parasol monoplane, fitted with semi-retractable tricycle landing gear. It carried a pilot, co-pilot, flight engineer, observer and radio operator, with provision for three pupils, and was also fitted with a flexible 7.7 mm machine gun in the bow, one in the dorsal position. Provision had been made to carry two 551 lb (250 kg) depth charges.

Three prototypes were built; the first had problems with handling and manoeuvrability, especially when alighting on the water. The following two prototypes had the 780-hp Nakajima Kotobuki 41 engines mounted lower on the wings, which increased the span and wing area and modified the flaps. With the modifications made, the aircraft were retested by the Navy and accepted. In 1942 the Aichi Navy Type 2 Training Flying Boat Model 11 (H9A1) was put into limited production, and between 1942 and 1943, the Aichi company built twenty-four and a further four were built by the Hikoki company. As the war progressed and Japan was forced on to the backfoot, the H9A1 was used for coastal patrols around Japan.

THE REWARDS OF PERSISTENCE

The needs of the Japanese Navy for accurate information derived from reconnaissance were becoming more and more demanding. Now the rewards for Japanese persistence with submarine aviation were to become apparent. A new monoplane reconnaissance floatplane, the Yokosuka E14Y1, to be codenamed the 'Glen' by the Allies, had been developed. This was a two-seat, twin-float monoplane that had a welded steel-tubular fuselage covered in a mixture of light alloy at its front section, and wood and fabric elsewhere. The wings were constructed of light metal spars with wooden ribs, and were covered in fabric. The monoplane was designed to be carried aboard a submarine in a hangar mounted on the deck. This was carried out by detaching the twin floats and their supporting struts, then removing the wings from the lower section of the fuselage at the spar fittings. The tail fin was detachable, so when the aircraft was slid back into its waterproof hangar on its launching trolley, there was sufficient clearance.

Powered by a 340-hp Hitachi Tempu 12, 9-cylinder radial engine, the Navy Type 0 Submarine-borne Reconnaissance Seaplane Model 11 (E14Y1), as it was designated, would perform reconnaissance tasks unimaginable a few years before, and would win a place in the annals of war. The first of the big Japanese submarines to have an aircraft hangar and catapult fitted to the front of the conning tower was the *I-9*. It had a length of 373 ft and a beam of 31 ft and carried one E14Y1 Glen.

The first operational missions flown by E14Y1s involved reconnaissance flights from the submarine *I-5* over Pearl Harbor in December 1941. The following month, the submarines *I-9* and *I-25* were equipped with this aircraft. Encouraged by the success of these sorties, the *I-25* moved into Australian waters where her E14Y1 made reconnaissance flights over Sydney Harbour on 18 February 1942, followed by a look at Melbourne on the 24th. Five days later the *I-25* launched her aircraft again to make a successful flight over Hobart, Tasmania, radioing back shipping information to the submarine. On the evening of 29 May, an E14Y1 from the *I-21* made a reconnaissance flight over Sydney Harbour. The information gained resulted in four midget submarines attacking two American warships – although unsuccessfully.

The *I-25* and *26* left Yokosuka in May 1942, to carry out an air reconnaissance of Kodiak in the Aleutian archipelago. Submarine-borne air reconnaissance was becoming more and more hazardous and on two occasions both the *I-26* and *I-25* had to crash-dive, only just managing to get their aircraft aboard and into their respective hangars. Later, off the Solomon Islands, the *I-9* lost her aircraft when she had to crash-dive, leaving her aircraft still on the deck. On 19 August 1942, another of the aircraft-carrying submarines, the *I-17*, was attacked on the surface while in the process of recovering her aircraft, and was sunk.

On 9 September 1942, an E14Y1 brought the war to the west coast of the United States, when Lt-Com. Meiji Tagami, commander of the *I-25* submarine, brought his boat to periscope depth six miles off Cape Blanco, Oregon. He called Warrant Flying Officer Nobuo Fujita, chief flying officer of the *I-25* to the conning tower and invited him to look through the periscope. The sea was now flat calm after ten days of heavy weather and the pilot, who was about to make history, saw the white face of Cape Blanco and its lighthouse flashing in the twilight.

On the morning of 9 September 1942, Warrant Officer Fujita and Petty Officer Shoji Okuda made their final preparations by placing some strands of hair, fingernail cuttings and a will in a special box made of paulownia wood. In the event of the aircraft not returning from the mission, these remains would be sent to their families.

The E14Y1 Glen, piloted by Warrant Flying Officer Fujita with his observer Petty Officer Okuda, was catapulted from the deck of the *I-25* submarine and headed towards Cape Blanco lighthouse. After crossing the coast, the aircraft swung north-east for the forests of Oregon. After flying for about fifty miles, Fujita ordered Okuda to drop the first bomb, which burst in a brilliant white light, setting fire to the forest below. Flying east for a few more miles, he ordered the second bomb to be dropped. Both bomb attacks caused fires. The aircraft then returned and rendezvoused with the submarine.

On the following day a second attack was made, again causing fires. On returning to the rendezvous point this time, Fujita discovered that the submarine was not in its pre-arranged position. After circling and searching for what must have felt like an interminable time, he spotted some oil on the surface of the sea and by following it came upon the submarine. By sheer luck, coupled with an oil leak from the submarine, he had managed to find the proverbial needle in the haystack. The aircraft was quickly taken aboard and stowed in its hangar, but it had been spotted flying away from the forests by civilian observers, who notified the authorities. US aircraft were scrambled, but sent in the wrong direction. A third raid was planned, but the weather had turned rough, so Lt-Com. Tagami called it off. This was the only drawback to using aircraft on submarines: they could only operate if the seas were calm, which restricted their usefulness to some degree.

On the journey home, they received a message from Tokyo saying that according to a radio broadcast in San Francisco, 'An aircraft, presumably from a submarine, had dropped incendiaries on the Oregon forest area, causing some casualties and some damage.' The sea was being scoured for the *I-25* by both aircraft and ships, so Lt-Com. Tagami kept the submarine submerged all day, only surfacing after dark to recharge her batteries. The *I-25* continued her patrol, and during the first half of October sank two oil tankers off the west coast and with her last torpedo sank a US submarine before returning to home waters. On his return to Yokosuka, Nobuo Fujita was hailed as a hero, his theory of carrying out bombing raids from submarines a proven fact. He was to see out the war as a flying instructor.

The only other recorded times that the American mainland was attacked during World War Two were in November and March 1944. The Imperial Japanese Army Special Balloon Regiment launched 9,300 thirty-five-foot-diameter hydrogen-filled paper balloons from eastern Honshu, Japan, to drift with the prevailing winds 6,000 miles across the Pacific. Many of the balloons were found between Alaska and Mexico (the exact number is not known) and they were armed with two small incendiary devices and a 35 lb anti-personnel bomb. The only known casualties were a woman and five children, who unwisely picked one up. It is interesting to note that this happened in Oregon, as did the first attack in 1942. The successful use by the Japanese of submarine-launched aircraft for reconnaissance and attack purposes in 1942 had amply proved the concept. Between 1941 and 1943 a total of 125 E14Y1 aircraft were built by the Watanabe company and one prototype was constructed by the Yokosuka Arsenal.

THE DEMANDS OF THE PACIFIC WAR (RECONNAISSANCE)

The increased fighting around the islands in the vast area of the Pacific Ocean meant that there were very few islands capable of sustaining an airfield, so the JNAF decided that floatplanes were needed for reconnaissance and patrol purposes. With this in mind, the JNAF issued a specification for a reconnaissance floatplane to the Aichi company. The prototype appeared in May 1942 and was of an-all metal construction with the exception of a wooden tailplane and wingtips, both of which were covered in fabric, as were the control surfaces. The Aichi AM-22, as it was initially called, was powered by a 1,300-hp Mitsubishi MK8A Kinsei 51 radial engine, giving it a maximum speed of 273 mph.

The twin metal floats were attached to the fuselage by inclined I-struts and to the wings by N-struts mounted vertically. For stowage aboard cruisers and seaplane tenders, the wings folded upwards. Hydraulically operated dive brakes were fitted to the front leg of the N-struts to enable the E16A1, as it was also known, to operate as a dive-bomber.

Armament consisted of two wing-mounted 7.7 mm machine guns and one flexible rear-firing 7.7 mm machine gun mounted in the rear cockpit.

The two other prototypes were used to test a variety of dive brakes and wing-mounted 20 mm cannons in place of the 7.7 mm machine guns. The production model had perforated dive brakes, strengthened floats and a greatly improved flap actuation mechanism. The first Aichi Navy Reconnaissance Seaplane (E16A1) Zuiun ('Auspicious Cloud'), known as 'Paul' to the Allies, came off the production line in August 1943. The early models were powered by the 1,300-hp Mitsubishi MK8A Kinsei 51 engine, later models were replaced with the 1,300-hp Mitsubishi MK8D Kinsei 54 and just before the Japanese capitulated the E16A1 was fitted with a 1,560-hp Mitsubishi MK8P Kinsei 62.

By the time the E16A1 entered service, the Japanese military had lost almost all of its air superiority and was fighting a defensive battle as Allied aircraft swarmed all over the Pacific skies. The E16A1 was no match for the American fighters, and the Japanese suffered heavy losses during the Philippines Campaign in 1944. At the very end what aircraft were left were used in kamikaze attacks.

A total of 256 E16A1s were built between 1942 and 1945 by the Aichi and Hikoki companies.

With the war in the Pacific escalating and the Japanese superiority in the air starting to wane as more and more American aircraft carriers appeared, the need for faster dive-bombers with a long range became a necessity. The JNAF issued a specification calling for a carrier-borne aircraft that would replace the Yokosuka D4Y and the Nakajima B6N. The requirements for the aircraft were onerous to say the least, as it had to be able to carry two 551 lb (250 kg) bombs or six 132 lb (60 kg) bombs or one 1,764 lb (800 kg) torpedo. Armament was to consist of two fixed forward-firing 20 mm cannons and one flexible rearward-firing 13 mm machine gun. It had to have a maximum speed of 354 mph (300 knots), a range of 1,151 miles (1,000 nautical miles) and have the manoeuvrability of a fighter. In addition to this, it had to have the capability of being stored aboard an aircraft carrier.

The Aichi company produced a proposal that covered all the requirements, and the design of the aircraft was unusual as it was an exceptionally large aircraft to be carried aboard an aircraft carrier. Powered by a 1,800-hp Nakajima Homare 11 18-cylinder radial engine, the Aichi B7A1 Ryusei (Shooting Star) (AM-23), as it was called, had an inverted gull-wing that enabled the length of wide-track undercarriage to be reduced to accommodate the four-bladed propeller's arc of 11 ft 5½ in (11 m).

The first prototype appeared in May 1942, and astounded everyone with its almost flawless display of its handling characteristics and manoeuvrability. There were some initial teething problems with the engine, and nine more prototypes were built using the same engine until all the problems had been sorted out. But it wasn't until April 1944 that a new engine became available – the 1,825-hp Nakajima Homare 12 – and the aircraft was put into production with the Aichi and Kokusho companies.

Production was slow and only 114 of the aircraft were built before the Aichi plant was destroyed, not by the Americans, but by a massive earthquake in May 1945, which brought production to an almost complete standstill. By this time it no longer mattered, as the Japanese Navy had lost their aircraft carriers and were now fighting a rearguard battle for survival. A handful of the B7A2s saw service operating from land bases, but made no impact at all.

The limited success of the B6N2 as a reconnaissance/dive-bomber aircraft prompted the Navy to approach the Nakajima company to produce a fast, long-range reconnaissance aircraft that could operate from the majority of the aircraft carriers. The specification laid down by the Navy included a top speed of 403 mph (350 kt) at 6,000 m; a climb rate of over 2,500 ft per minute; a maximum range of 3,078 miles (3,500 nautical miles); and a landing speed of less than 81 mph (70 kt).

The specifications were onerous to say the least, and to meet them the engineers and designers realised that they would have to find a way of minimising the drag and find engines powerful enough to meet the requirements. Initially they considered a revolutionary idea of installing two engines inside the fuselage and by means of an extension gear system, powering two propellers on the leading edge of each wing. This was considered to be too high-maintenance with the potential to cause numerous problems, so it was decided to install the latest Nakajima engine, the 1,820-hp Nakajima NK9B Homare 11 18-cylinder, air-cooled, radial, which drove a metal four-bladed propeller. This new engine was much smaller in diameter, enabling the designers to come up with a slimmer fuselage. The oil cooler was mounted offset to port and in an external position. By canting the vertical tail surfaces in a forward position, it kept the fuselage within the size required for the deck elevators on the aircraft carriers.

The wing area, although constrained because of the restrictions of size aboard the carriers, was still able to contain four protected and two unprotected fuel tanks containing a total of 299 gallons (1,360 litres). A combination of Fowler and split flaps were fitted to the trailing edges of the wings as well as slats to the leading edges, which enabled the landing speed of the aircraft to attain the 70 kt landing speed required.

The long, glazed cockpit, which contained a crew of three, a pilot, a navigator/observer and a radio operator/gunner, stretched two-thirds the length of the fuselage. The only armament carried was a single, flexible, rear-firing 7.92 mm machine gun.

The Nakajima C6N1 Saiun prototype, as it was designated, made its first appearance in May 1943. All the tests were satisfactory as far as the handling characteristics and manoeuvrability were concerned, but there were numerous teething problems with the Homare engine. When subjected to flight tests by the JNAF it failed to meet the necessary speed requirements, and in an effort to meet these, the Nakajima company built a total of nineteen prototype and pre-production aircraft. The more powerful NK9H Homare 21 engine was fitted, but still the required speed eluded the C6N1.

Despite this setback, it was still one of the fastest reconnaissance aircraft around and the Navy accepted this. The aircraft was put into production. During the battle of the Marianas the C6N1 Saiun was fitted with a torpedo-shaped ventral drop fuel tank carrying 160 gallonss (730 litres), and because its speed was as fast as the F6F Hellcat of the US Navy, it was able to shadow the US Pacific Fleet with a high degree of immunity.

This ability brought it to the attention of the Allies and it was given the code name 'Myrt'. At one point it was decided to convert a number of the C6N1 to carrier-borne attack bombers with provision for carrying a torpedo and possessing forward-firing machine guns, but the loss of almost all of the aircraft carriers forced this idea to be shelved. However it discovered a new role when the Boeing B-29s started carrying out intense bombing raids against the Japanese mainland. A number of the C6N1s were converted to night fighters by removing the radio operator/gunner and his equipment and installing a pair of 20 mm cannons that were mounted obliquely in the fuselage. The C6N1-Ss, as they were called, were now able to attain the original speed requirements because of the reduction in weight, and became the fastest night fighters in Japan's armoury.

A number of other models were proposed, but the war ended while they were still on the drawing boards. The C6N1's last claim to fame was that it was shot down at 05.40 hours on 15 August 1945, just five minutes before the war was officially declared to be over, the last confirmed aerial victory of the Pacific War. Between March 1943 and August 1945, a total of 463 Nakajima C6N1s had been built.

Late in 1940, the JNAF put forward a requirement for a single-engined crew trainer to replace their Mitsubishi K3M. The Watanabe company put forward a proposal for an all-metal monoplane trainer with the wing mounted mid-fuselage and fabric-covered control surfaces, retractable undercarriage with a deep belly fuselage and the pilot and radio operator/gunner seated in tandem above the level of the wing, and the instructor, navigator and bombardier seated in a cabin area under the wing.

A prototype appeared in November 1942, powered by a 515-hp Hitachi GK2B *Amakaze* 21, 9-cylinder air-cooled radial engine, and it was given the designation K11W. The initial test flights went well and the aircraft was sent to the Navy for further testing and evaluation. The Watanabe company was reorganised and became the Kyushu Hikoki KK at the end of 1941 and was given the contract to build the new trainer. The first of the new aircraft, given the official designation of Navy Operations Trainer Shiragiku ('White Chrysanthemum') Model 11 (K11W), went into service in July 1943. It was armed with a single flexible rear-firing 7.7 mm machine gun and facilities to carry either two 66 lb (30 kg) bombs or one 551 lb (250 kg) bomb.

An all-wood version was built, to be used for transport and anti-submarine patrols, but they were not successful as they were discovered to be underpowered. A total of 798 K11W1 training aircraft were built, with a very small number of the wooden version, known as the K11W2, included.

The building of Western aircraft under licence was still a priority with Japanese aircraft manufacturers. At the end of 1938, Kawasaki was about to build a copy of the Lockheed 14-WG3 twin-engined transport when the Koku Hombu told the company to build an improved version of the aircraft for military use with the designation Ki-56.

Using the Lockheed aircraft as a template, the Kawasaki engineers lengthened the fuselage by 4 ft 11 in. (1.5 m), redesigned the Fowler flaps to improve their efficiency and reduced the weight of the wing structure. The engines that had been on the original Lockheed copy, the Mitsubishi 900-hp Army Type 99 Model radials, were replaced by two Nakajima Ha-25 950-hp Army Type 99 radials. A large cargo-loading door was cut into the port side of the fuselage that also incorporated a smaller crew entry door. Despite the increased overall size of the Kawasaki Ki-56, as it had been designated, it was 115 lb (52 kg) lighter than the Lockheed 14-WG3.

Two prototypes were completed in November 1940 and both aircraft were immediately subjected to flight trials, such was the urgency of the requirement. The overall performance of the aircraft was better than the Lockheed and the Ki-56 was put into production, and between 1941 and 1943 a total of 121 Ki-56 aircraft were built.

THE ARMY'S ZERO

Next to the Zeke fighter, and considered by some to be even better, was the Nakajima Ki-84 Hayate. The JAAF was desperate for an all-purpose, long-range fighter that had a top speed of around 400 mph (680 km/h), a combat operating time of 1½ hours at around 250 miles (400 km) from its base. Armament was to consist of two engine-mounted 12.7 mm machine guns and two wing-mounted 20 mm cannons. The aircraft also had to have armour plating for the crew and be fitted with self-sealing fuel tanks.

The design of the fighter started at the beginning of 1942, and in less than a year a prototype was produced. Of a conventional design, the low-wing monoplane was powered by a 1,800-hp Nakajima Ha-45 18-cylinder, air-cooled radial engine. A centre-line fuel tank was mounted beneath the fuselage, which was ejected by means of powerful springs when exhausted.

In April 1943 the Ki-84 took off on its maiden flight and performed without a problem. Two months later a second prototype came off the production line, the first of eighty-three Ki-84s to be used during service trials. Some of these were handed over to the JAAF for test and evaluation by the Army test pilots. Without exception, all those who flew the aircraft were delighted with its performance and manoeuvrability. They were also pleased with its climb rate of 2,700 feet per minute, which was quite exceptional for the time.

The tests completed, this was the first time that any Japanese-built aircraft was ready for production without some major modifications being necessary. There were some minor modifications such as the shape and area of the fin and rudder being altered to improve control on take-off, and bomb racks were fitted on some of the aircraft, enabling them to carry 551 lb (250 kg) bombs.

One of the first units to fly the Ki-84 was an experimental Chutai, which carried out extensive testing under combat conditions. The results came back and the Ministry of Munitions, who were the body that placed all production orders, immediately instructed the Nakajima company to put the aircraft into production with the designation of Army Type 4 Fighter Model 1A Hayate (Ki-84-1a).

Once in production, the aircraft were produced at the rate of 250 per month and the first unit to receive the aircraft was the Twenty-Second Sentai, most of whose personnel had been with the now-defunct experimental Chutai who had conducted the combat trials. Their first combat missions were carried out in China against the US Fourteenth Air Force, and were so successful that they were transferred to the Philippines where they came up against some of the Allies' best fighters. Given the Allied codename 'Frank', the Ki-84 distinguished itself and earned the respect of the Allied pilots.

The Japanese forces were now on the defensive, and the strain placed upon the fighter units was increasing by the day. Because of the extensive pressure the aircraft were being subjected to, maintenance schedules were being dispensed with, resulting in problems in the engines and hydraulic systems. The number of aircraft being lost either to Allied fighters, or crashes because of hydraulic failure affecting the landing gear, was putting increased pressure on the manufacturers.

The factories in Japan were doing their best to cope with shortages of raw materials and the persistent day and night bombings by the Americans. The rapidly dwindling supply of aluminium resulted in the rear fuselage section of the Ki-84-II being made out of wood. This model was powered by the latest 1,800-hp Nakajima Ha-45 engines, which had a low-pressure fuel injection. Nakajima were having desperate problems because of the lack of aluminium, so much so in fact that they told the Ministry of Munitions that if they were to continue using wood to a greater extent, there would have to be a complete re-design of the Ki-84's airframe.

It was decided to make an all-wood model, and the task was given to the Tachikawa company to design an all-wood version of the Ki-84. The result was an all-wood airframe designated the Ki-106, which was built by the Prince Aircraft Company (Ohiji Koku KK). The external finish to the plywood skin was achieved by applying a thick coat of lacquer. The first test flight of the Ki-106, powered by a 1,990-hp Nakajima air-cooled, radial engine, showed that the bonding of the wood had to be strengthened considerably. This done, all the tests were carried out satisfactorily, but the war ended before the aircraft could be put into production. At the end of the war a number of variants were developed but none were ever put into production. However, the total number of Ki-84 aircraft built was 3,514, one of the highest numbers of aircraft ever built by a Japanese company.

A SPECIALIST KAMIKAZE AIRCRAFT

One of the most unusual aircraft built by the Japanese was the Nakajima Ki-115 Tsurugi ('Sabre'). It was never designed for combat purposes, but for kamikaze missions. The development of the aircraft was simplicity itself because it was designed and built for just one mission and one mission only. The only provision required by the aircraft was recessed crutch attachment for a single 1,764 lb (800 kg) bomb.

The fuselage had an open cockpit and was partially covered at the front section by an all-metal skin, whilst the rear section of the fuselage was fabric covered. The wings were mounted low on the fuselage and covered with a stressed metal, whilst the tail section was constructed of wood and covered with fabric. The undercarriage was non-retractable and jettisoned after take-off.

The aircraft was powered by a 1,150-hp Nakajima 23 Ha-35, 14-cylinder, air-cooled, radial engine that gave the aircraft a top speed of 320 mph (515 km/h) after jettisoning the undercarriage. The engine was fitted to the fuselage by just four bolts. The construction of the aircraft was carried out by semi-skilled labour, which showed itself during the first test flight. The handling characteristics on the ground were very poor, mainly because of the poor vision from the cockpit. The undercarriage, which was constructed from welded steel piping and had no shock absorbers, had to be replaced with a much stronger one with shock absorbers. In addition auxiliary flaps were fitted to the inboard wing trailing edges. In an effort to boost the aircraft's final dive, two solid fuel rockets were mounted beneath each wing.

The aircraft had been designed to be flown by a pilot with only the most basic flight training. The vast majority of these young pilots were barely into their teens, and were to find themselves facing certain death after just a few weeks of training.

With the increasing push forward by the Allies on the ground being supported by ever-increasing numbers of fighter aircraft, the Japanese Army waited impatiently for the arrival of their new twin-engined heavy fighter, the Kawasaki Ki-96. The prototype was nearing completion when the designer of the aircraft approached the Army with a suggestion that they build a ground attack version. The Army agreed and work started on a prototype in August 1943.

The designer, Takeo Doi, decided to use the airframe and engines that were being used on the Ki-96 prototype, but added additional armour protection for the crew and for the fuel tanks. Armament was also increased, in the form of a nose-mounted 57 mm cannon, two fuselage-mounted 20 mm cannons and one flexible rear-mounted 12.7 mm machine gun. The first of three prototypes to be built was powered by two 1,500-hp Mitsubishi Ha-112-II, 14-cylinder, air-cooled radial engines and took to the air in March 1944.

In quick succession there followed a further two prototypes and twenty pre-production Kawasaki Ki-102 aircraft, as it had been designated, for service trials. The trials were quickly carried out and the aircraft displayed excellent handling characteristics and performance figures. The only concern was that the Ki-102 showed marked directional instability on landing approach. This problem was quickly resolved and the aircraft was immediately put into production as the Army Type 4 Assault Plane (Ki-102b).

The war at this time was not going well for the Japanese; they were now on the defensive, trying desperately to stem the advancing Allied forces as they swept across the Pacific. The majority of the Ki-102bs as they came off the production line were held in reserve in Japan, but a small number were involved during the Okinawa campaign. It was there that the Allies first saw the aircraft, and gave it the codename 'Randy'.

The JAAF was still waiting for the Ki-96 to be completed, and put forward a proposal that it would like a high-altitude fighter. It was decided to shelve the Ki-96 and concentrate on another version, the Ki-108, but time was running out and the Koku Hombu ordered that the Ki-102 be modified and used as a high-altitude fighter. A number of modifications were made, including the fitting of two 1,500-hp Mitsubishi Ha-112-II Ru engines fitted with turbosuperchargers, and additional firepower. Only fifteen were ever built and delivered to the Army, but they achieved very little as there were ongoing problems with the turbosuperchargers.

THE NEED FOR A NIGHT FIGHTER – BOMBERS AND INTERCEPTORS

With the bombing war now being carried to Japan itself by the high-flying American B-29 bombers, the need for a specialised night fighter became more and more desperate. The JAAF had one converted night fighter, the Kawasaki Ki-45 Toryu, but was desperate for a purpose-built one. Work immediately started on converting some of the Ki-102bs by lengthening the fuselages and redesigning the cockpits and tail surfaces. The wingspan and area were increased and a revolving radar dish was installed beneath a Plexiglas radome and mounted on top of the fuselage. Two 30 mm cannons were mounted obliquely in the belly of the fuselage and two 20 mm cannons were mounted obliquely in the fuselage behind the cockpit. Trials were commenced in July 1945, but the end of the war put an end to the production of the aircraft. Between 1944 and 1945 a total of 288 Ki-102s were built by the Kawasaki company.

At the beginning of the Pacific War the Japanese were in almost total control of the skies in the Pacific, but as the war progressed the Americans began to re-build their war machine and B-17 bombers began to hit back. The arrival of the B-29 Superfortress and its ability to fly at heights unattainable by Japanese fighter aircraft caused the Koku Hombu to look towards developing a high-altitude interceptor. In an effort to prevent these high-flying bombers, the Japanese approached the Mitsubishi company for an answer. At the time, the latest bomber, the Mitsubishi Ki-67, was undergoing flight trials and proving to be a very fast and extremely manoeuvrable aircraft, despite its size and weight.

It was decided to use the aircraft as a basis for a hunter/killer team, which was to be designated the Mitsubishi Ki-109. The hunter, Ki-109b, was to be equipped with radar and a 40 cm searchlight, while the killer, Ki-109a, was to have two oblique-firing 37 mm cannons mounted in the rear section of the fuselage. One later modification was the fitting of a 75 mm Type 88 cannon in the nose of the Ki-109a, which it was hoped would enable the aircraft to fire upon the B-17s and B-29s without coming within range of the enemy's guns.

The first prototype appeared in August 1944 and was identical to the Ki-67, retaining all the gun positions in addition to the large, nose-mounted 75 mm cannon. The tests were so successful that the Army placed an immediate order for forty-four of the aircraft. The first twenty-two of the production line were powered by two 1,900-hp Mitsubishi Ha 104 air-cooled radial engines. An attempt to fit the remaining part of the order with Ha-104 Ru engines that had been fitted with the Ru-3 exhaust-driven turbo superchargers was shelved when the second prototype's performance proved to be ineffective. A third prototype was produced using a solid propellant rocket battery to try to improve the climb rate, but again this was ineffective.

A fourth prototype had the gun blisters on either side of the fuselage and the dorsal turret removed, together with the bomb bay. The Type 88 75 mm cannon that was mounted in the nose was hand-loaded by the co-pilot and only fifteen shells were carried. Still the Ki-109 lacked the high-altitude performance needed to intercept the B-29s, but by this time the Americans had gained control of the skies and were now carrying out low-level, night-bombing raids with a fighter escort.

In January 1940, with the Sino-Japanese conflict still going on, the JAAF saw the need for a heavy bomber as it prepared to engage the Soviet Union along the Manchukuo-Siberian border. In the February it issued a specification to the Mitsubishi company for a tactical heavy bomber that had to have the following requirements: a maximum range of 2,360 miles (3,800 km); a maximum speed of 342 mph (550 km/h); a maximum bomb load of eight 220 lb(100 kg) or three 551 lb(250 kg) or one 1,102 lb(500 kg); and the ability to carry a crew of between six to eight at an operating altitude between 13,125 ft (4,000 m) and 22,965 ft (7,000 m). It was to have a defensive armament of one 7.7 mm machine gun in the nose and port and starboard blisters, and one 12.7 mm machine gun in the dorsal and tail turret. The aircraft was to be powered by two 1,450-hp Mitsubishi Ha-101 radials, two 1,870-hp Nakajima Ha-103 radials, or two 1,900-hp Mitsubishi Ha-104 radials.

Using the design of the successful Mitsubishi G4M1 bomber as a basis, Chief Engineer Ozawa concentrated his efforts on the ease of production by constructing sub-assemblies and ensuring that all the fuel and oil tanks were of the self-sealing type with armour protection. By December 1942 the three prototypes requested by the Koku Hombu had been completed and were ready for testing. The Army then requested that additional prototypes and service trials aircraft be built, which of course held the programme back somewhat.

The first prototype took to the air on 27 December 1942 and despite some excessive control sensitivity, longitudinal stability and a speed that fell slightly below the requirement, the Mitsubishi Ki-67 Hiryu ('Flying Dragon') exceeded all expectations. So pleased with the results were the Army that they ordered that two of the service trials aircraft were to be modified to carry torpedoes. They were then taken to the Yokosuka Naval Air Station for trials. So successful were the trials that the Navy ordered 100 of the aircraft to be built as torpedo-bombers.

The Army continued to look upon the Ki-67 with great expectations, and continually planned new roles for the aircraft by adding new equipment. The development of the Ki-67 finally reached a stage when its production was in jeopardy because no one would make the decision as to what would be the standard configuration of the aircraft in terms of production.

On 2 December 1943, the Koku Hombu stepped in and ordered a stop to the various designs and modifications, and told the Mitsubishi company to produce a single version of the aircraft as laid down in the original requirements. It wasn't until October 1944 that the Ki-67 first saw action, when it was used as a torpedo-bomber off Formosa and during the American landings on Okinawa. The heavy bomber version operated from bases in China and Hamamatsu, making repeated attacks against airfields in the Marianas used by Boeing B-29 bombers.

In Japan all the stops were pulled out in the production of the Ki-67, but the intensive bombings by the Americans and an earthquake in December 1944 caused production of the aircraft to be slowed dramatically. By the end of the war only 698 Ki-67 heavy bombers had been built. The bomber appeared too late for the Japanese, as the majority were crewed by inexperienced pilots straight out of training school, who were being thrown against battle-hardened Allied pilots.

The Mitsubishi company had been looking at producing a single-seat interceptor fighter based on the Navy's requirement for an aircraft that had a high rate of climb and speed. The specifications required by the Navy were for the aircraft to have a maximum speed of 373 mph (600 km/h), a climb rate of 1,000 feet per 1½ minutes, an operating ceiling of 19,685 ft (6,000 m), a landing speed not more than 81 mph (130 km/h), a take-off distance of 984 ft (300 m) and an endurance of at least 45 minutes.

Chief designer Jiro Horikoshi decided that the best engine for the aircraft, designated the Mitsubishi J2M1 Raiden, would be a 1,430-hp Mitsubishi Kasei 14-cylinder radial, with an air-driven fan that would allow the fitting of a fully tapered cowling. In an effort to reduce the drag, the cockpit canopy and windscreen were very shallow, and the fitting of combat flaps was to improve the aircraft's manoeuvrability.

Because of the urgent need to develop the A6M series of aircraft that was already in production, and because of the problems with the engine cooling system of the new aircraft, it wasn't until the beginning of 1942 that the first prototype appeared. The first trials were not the success hoped for: Navy test pilots complained bitterly about the extremely poor visibility afforded them by the sloping windscreen, and it was discovered that the undercarriage would not retract at speeds over 100 mph. There were other problems, like the propeller pitch change mechanism, which was totally unreliable, with the result that the aircraft's speed and rate of climb were well below what was laid down in the specifications.

The Navy sent the aircraft back to Mitsubishi with instructions to sort the problems out. The first thing that was replaced was the cockpit canopy, and a conventional bulletproof flat screen replaced the windshield. A new engine was installed – a 1,575-hp Mitsubishi MK4R-A Kasei 23a 14-cylinder air-cooled radial with a four-bladed constant speed propeller. The J2M2, as it was designated, was accepted and went into production in October 1942. Problems raised their head once again when engine vibration caused the aircraft to be almost uncontrollable. Improved engine mountings and modifications to the propeller eliminated this.

The aircraft went back into production but because of other commitments and minor problems with the J2M2, delivery was slow. This was not helped by two accidents that occurred after delivery had been made to one of the squadrons. The first was just after the aircraft took off; it suddenly went into a dive, killing the pilot as it smashed into the ground. The Mitsubishi engineers examined what was left of the wreckage but could find nothing that would have caused the aircraft to go into a dive, and put it down to pilot error. The second happened under similar conditions, but this time the pilot managed to regain some semblance of control and crash-landed the aircraft. On examining the aircraft the engineers found that the tail wheel strut, after being retracted, pressed against the torque tube lever, jamming the controls into a dive position. The tail wheel struts were all immediately modified on the other J2M2 aircraft, which resolved the problem.

Teething problems with the engines seemed to dog the J2M2 and the next version, the J2M3. The later version only differed from the others by more powerful armament. A radical new approach with the next two variants, the J2M4 and J2M5, was the most ambitious attempt to get a high-altitude fighter operational. The J2M4 was fitted with a new engine, the Mitsubishi turbo supercharged MK4R-C Kasei 23c engine. This made the aircraft capable of speeds around 362 mph at 30,185 ft, which would have made it the ideal aircraft to intercept the high-flying Boeing B-29 bombers, which flew almost unchallenged. Only two of the prototypes were built, as problems with the turbosupercharger caused the project to be cancelled.

However, the J2M5 was fitted with a Mitsubishi MK4U Kasei 26c engine with a three-stage supercharger that was driven mechanically. This proved to be a success, and came at a time when B-29s were beginning to bomb Japan on a regular basis. The aircraft was immediately rushed into production at the Koza Naval Air Arsenal, which built thirty-four J2M5s. In total 476 of the J2M models were built, and although they suffered a large number of problems the Raiden was one of Japan's most successful fighter aircraft. It is fortunate for the Allies that these problems prevented the aircraft's mass production, as there is no doubt that it would have caused serious problems.

In December 1941, the Navy ordered twenty-seven G4M1s to be sent to bases in Indochina in preparation for the attack on the British battleships there. In Formosa, ninety-three attack bombers were prepared for attacks on the American forces in the Philippines. Within days of the Japanese attack on Pearl Harbor, the Mitsubishi G4M1s had attacked and sunk British battleships.

Up to this point the G4M1s had faced almost no aerial opposition, but as the Allies recovered from the shock, their aircraft numbers were increased and they started to fight back. It was then that the G4M1 losses started to mount because of the lack of armour plating for the crews and the fuel tanks. Because of the rapid advances made by the Japanese, attacks were being made on more distant targets like Port Moresby and Australia. They met fierce resistance from Allied fighters stationed there and suffered severe losses. In an effort to reverse the situation, Mitsubishi developed a variation on the G4M1, the G4M2, also known as the Navy Type 1 Attack Bomber Model 12.

Powered by twin Mitsubishi 1,825-hp Kasei 14-cylinder, air-cooled radial engines, the G4M2 had the gun blisters on the fuselage removed, and the tail turret was altered to house a 20 mm cannon. In an attempt to protect the wing fuel tanks and the fuselage tanks, rubber sheeting was used. The modifications increased the weight considerably with the result that the performance deteriorated. The first of the new aircraft appeared in November 1942.

The prototype was powered by two 1,800-hp Mitsubishi MK4P *Kasei* 21 radial engines. The G4M2 had a larger tailplane area and rounded wing and tail tips to help increase stability, a greenhouse-type nose section with gun ports on either side, a 20 mm cannon in a hydraulically-operated dorsal turret and the installation of an auxiliary fuel tank in the fuselage.

A second prototype was produced almost identical to the first, but the third model was tested with bombs being carried in a bomb bay. Tests were rapidly carried out because of the escalating war and the sudden halt of the Japanese advance. It was now becoming apparent that the Allies were gaining in strength and the superior weight in numbers of both men and machines was beginning to take effect.

The G4M2 was put into production alongside the G4M1, but as the new Kasei 21 engines became available, the production line of the G4M1 was stopped and replaced with an additional G4M2 line. Other versions appeared: the G4M2a Model 24A, 24B and 24C, all of which had different types of armament, although the later version of the 24C was fitted with air-to-surface radar.

Slowly but surely, the G4M2 replaced the G4M1 as a front-line aircraft, and by the end of 1944, all had been replaced. The tide of war was turning, however, and as the Allied forces gained in strength, the losses of the Japanese aircraft mounted daily. In an effort to stem the tide, the Japanese turned to using a piloted missile called the Ohka. A number of the G4M2a Model 24B and C had their bomb-bay doors removed and a clamp system fitted that enabled them to carry one of these rocket-propelled suicide missiles. The problem was that the additional weight, which was considerable, reduced the performance of the mother aircraft dramatically. On the first mission, sixteen G4M2as took off, but were so slow and ponderous that all were shot down before they got anywhere near their intended target. This was to be repeated time and time again, although occasionally the odd one did get through.

Another version appeared towards the end of 1944, Model 34, which had improved crew protection and a modified tail turret, but it was all too late and they were still undergoing flight trials when the war ended. Just four days after the war was officially declared over, two all-white G4M1s with green crosses painted on the fuselage and the tail, and escorted by Boeing B-17H bombers, carried the Japanese surrender delegation to Ie Shima. Between 1939 and 1945, a total of 2,446 G4M and G6Ms were built, the largest number of Japanese bombers ever.

THE OHKA SUICIDE WEAPON

By the summer of 1944, the Japanese High Command realised that they were now fighting for survival, not for victory. Even the most fanatical of them knew that the Allies were at the point of overwhelming their beleaguered forces, and they turned to desperate measures. One of these was put forward by a Naval officer, Ensign Mitsuo Ohta, who designed a rocket-propelled suicide aircraft. His design was presented to the Navy at Yokosuka, who immediately accepted the proposal and ordered engineers to draw up detailed drawings.

It was initially designed as an anti-invasion weapon to be launched from a mother aircraft. It would glide after being released, then fire its three solid-propellant rockets, either singly or in unison, and accelerate towards its target. Because the aircraft was on a one-way mission, the instrumentation was kept to the barest minimum as the pilots who flew them would have had only the basic flying instruction.

The design of the aircraft was such that it was built of wood and non-critical metal and could be assembled by unskilled labour. The only serious requirement was that it had to have very good manoeuvrability, as this would enable the pilot to achieve reasonable accuracy. The first of the aircraft came off the production line only weeks after the design stage, and by the end of September 1944 ten of the aircraft were ready.

Given the designation Navy Suicide Attacker Ohka ('Cherry Blossom') Model 11 (MXY7), it was equipped with a 2,646 lb (1,200 kg) warhead and was carried in the bomb bay of a specially modified Mitsubishi G4M2. At the beginning of October, the initial flight trials were carried out without power, and it wasn't until the middle of November that the first rocket-powered flights were carried out.

Such was the urgency that the Navy ordered full production to start without even getting the full results of the trials. In the next six months a total of 755 Ohka Model 11 aircraft were built. It wasn't until 21 March 1945 that the Ohka Model 11 first went into action. Sixteen Mitsubishi G4M2s took off, each with an Ohka Model 11 tucked beneath their fuselages, but the slow and cumbersome bomber was an easy target for the Allied intercept fighters and the Ohkas were released prematurely, and none of them reached their targets.

The first success, however, was a major one, when on 1 April 1945 a number of the Ohka were released, causing severe damage to the American battleship USS *West Virginia*, and three transport ships. On 12 April the destroyer USS *Mannert L. Abele* was sunk off Okinawa; seventy-nine crewmen were killed or missing and thirty-five were wounded after the ship broke in half. It soon became obvious, however, that the Mitsubishi G4M2 parent aircraft was far too slow, and when approaching within a few miles of heavily defended targets they suddenly found themselves the targets.

Because of this, production of the Ohka 11 was ceased and an improved version, the Ohka Model 22, was developed. In the meantime, the Ohka K-1 unpowered version appeared with ballast replacing the warhead, and with retractable skids fitted. This model was used for training purposes and the water ballast was released during the landing approach to reduce the landing speed to 138 mph (120 knots). Forty-five K-1s were built during 1945.

The Model 22 had a smaller wingspan than the Model 11, because it was to be carried beneath the Yokosuka P1Y1 and there was a limited clearance beneath the aircraft's fuselage. The three rockets were replaced by a Tsu-11 Campini-type jet engine that had a 100-hp Hitachi 4-cylinder in-line engine for a gas generator. This improved power plant enabled the Ohka Model 22 to be dropped much farther from the intended targets, enhancing the element of surprise. Production started in the middle of 1945, and fifty of the aircraft were built. The Aichi company were then tasked with carrying on the production, but badly bomb-damaged factories forced the high command to consider building the aircraft in underground bunkers. The war ended before the bunkers could be built.

A number of other variations were on the drawing boards at the time of surrender, including an enlarged version of the Model 22 (the Model 33) and the Model 43A, which was designed to be launched from submarines. A total of 852 Ohka suicide aircraft were built

Another aircraft under development at the time was the single-engined dive-bomber, the Yokosuka D4Y Suisei ('Comet'). The Japanese Navy had acquired the production rights of the Heinkel He 118V4 in the spring of 1938, and based the design of the D4Y, albeit a smaller version, on this model. The designer and chief engineer, Masao Yamana, produced a two-seat, mid-wing monoplane powered by a 960-hp Daimler-Benz DB 600G engine. It also featured an internal bomb bay designed to carry a single 500 kg bomb. Beneath the wings were fitted three electrically operated dive brakes, just in front of the landing flaps. The undercarriage retracted into the wings just in front of the main spar.

The Navy had laid down requirements for the aircraft, and they included that it should have a maximum speed of 322 mph (280 kts), a cruising speed of 265 mph (230 kts), a range of 921 miles (800 nautical miles) and should be versatile enough to operate from both land bases and aircraft carriers. The armament was to consist of two forward-firing 7.7 mm machine guns mounted in the upper section of the fuselage and a single flexible rear-firing 7.92 mm machine gun.

The prototype made its maiden flight in December 1940 and exceeded all the Navy's expectations. Four more prototypes followed, all with minor modifications and upgrades, and all exceeded what was expected of them. But a serious problem arose when it was put through its dive tests as violent wing flutter occurred, resulting in cracks appearing in the wing spars. Production plans were immediately halted.

After extensive examination it was decided to reinforce the wing spars and fit larger dive brakes, but it wasn't until March 1943 that the D4Y1 Suisei was accepted by the Navy as a dive-bomber. Demand for the aircraft was speeded up as Japanese aircraft losses mounted and the Allied push gathered momentum. In June 1944, the First, Second and Third Koku Sentais (Carrier Divisions) with 141 D4Y1 Suiseis, together with a number of other aircraft, were embarked aboard nine aircraft carriers in an effort to halt the US invasion of the Marianas. Almost all were shot down in what has since become to be known as the 'Marianas Turkey Shoot', and not one American aircraft carrier was sunk. The Japanese lost over 240 aircraft and two aircraft carriers: the *Shokaku* and the *Taiho*. The tide had turned.

The next version of the D4Y1, the D4Y2, was well underway on the production line. Powered by a 1,400-hp Aichi Atsuta 32 engine, the D4Y2 was known as the Suisei Carrier Bomber Model 12 and was almost identical to the D4Y1. The only real differences were the Aichi engine and a flexible, rear-firing, 13 mm machine gun in place of the 7.92 mm machine gun. This version was also fitted with catapult equipment, enabling it to be launched from much smaller, converted aircraft carriers and was known as the D4Y2a KAI Model 22A.

By this time Allied aircraft like the Grumman Hellcat and Vought Corsair had helped re-establish control over the Pacific, and during the battle for the Philippines, the Japanese lost significant numbers of aircraft and pilots. A number of Suiseis were relegated to kamikaze missions. The fitting of the Atsuta engine was initially well received, but as it was put under battle conditions it became plagued with problems and maintenance under these conditions created many difficulties. Some of the senior officers demanded that the engine be replaced, so a 1,560-hp Mitsubishi MK8P Kinsei 62, 14-cylinder radial was installed. Because it was somewhat larger than the Atsuta, the pilot's visibility was slightly reduced during take-offs and landings from a carrier, but it proved to be a more reliable engine and was accepted without reservation.

A small number of variations appeared over the following months, including a specialized suicide bomber. This meant that it was designed as a single-seat fighter, carrying an 800 kg bomb that was semi-recessed beneath the fuselage. It was also fitted with three auxiliary rockets that could be used for boosting take-offs from small landing strips, or to increase speed during its suicide dive. It was known as the Suisei Special Attack Bomber Model 43.

Other versions were still under development when the war ended, and a total of 2,038 D4Ys were built between 1942 and 1945. These were made up as follows: 860 D4Y1s; 336 D4Y2s; 536 D4Y3s; and 300 D4Y4s

With the development of A6M carrier aircraft, now being produced in greater quantities, the Navy had not been idle. Their part in the Shanghai Incidents boosted their proposal for more land-based squadrons and the building of two more aircraft carriers of 20,000 tons, each capable of carrying seventy-six aircraft, and three seaplane carriers of 13,000 tons, capable of carrying thirty-six aircraft each.

In between the development of the fighters, reconnaissance aircraft and bombers came the need for personnel transport aircraft. Although they didn't have the same priority level as the other military aircraft, they were a gap in the necessary part of the Army's war machine. In an effort to close this gap, the Koku Hombu selected a twin-engined airliner that had been developed for Japan Air Lines by Mitsubishi from their Ki-21 heavy bomber.

The Army issued a specification that required a transport aircraft that could carry at least eleven passengers plus a crew of four, and 661 lb (300 kg) of freight over a distance of 870 miles (1,400 km) at a cruising speed of 186 mph (300 km/h), and at a height of 6,560 ft (4,000 m).

Taking the design of the Ki-21, Mitsubishi engineers retained the wings, tail section, undercarriage, and cockpit section and just replaced the fuselage. Inside the new fuselage the passengers were accommodated in two rows of single seats with room at the rear of the aircraft for freight. The wings of the Mitsubishi Ki-57, as it was designated, were mounted low on the fuselage, unlike the wings of its predecessor the Ki-21, which were mounted mid-fuselage.

At the beginning of 1941, a small number of Ki-57s were passed to the Navy to be used either as a communications or logistic aircraft, or as paratrooper carriers. During the Japanese paratrooper attack on the oil refineries and airfield at Palambang, southern Sumatra, it was the Ki-57 that carried the paratroopers.

An improved version of the Ki-57-I, the Ki-57-II appeared in May 1942, powered by two 1,080-hp Mitsubishi Ha-102 radial engines. The engine nacelles had to be re-designed to accommodate them and there were a few very minor changes. The Ki-57 was a very unassuming aircraft, but throughout the war period it was a veritable workhorse and played its part in the conflict.

A FLOATPLANE FIGHTER

The need to give air support to ground troops where no airfield was available prompted the Navy to commission the Kawanishi company to produce a singe-seat, single-engined floatplane fighter. This resulted in the Kawanishi N1K1 Kyofu, powered by a 1,460-hp

Mitsubishi Kasei 14-cylinder engine, which in turn powered two contra-rotating two-blade propellers. This was to offset the anticipated torque that would develop on take-off. It was fitted with a central float, which was mounted directly beneath the fuselage and supported by a V-shaped strut at the front and an I-shaped strut at its rear. Initially it had been proposed that the two outrigger floats at the end of each wing would be retractable, but difficulties had been experienced with this design on an earlier model, so it was decided to mount them in the fixed position.

After a series of trials, problems with the contra-rotating propellers caused them to be removed and a conventional three-bladed propeller was installed on a new engine. The powerful torque created by this engine required a very skilled pilot to handle the aircraft while landing and taking off, but once in the air the addition of combat flaps made the aircraft extremely manoeuvrable. Two wing-mounted 20 mm cannons and two fuselage-mounted 7.7 mm machine guns made the N1K1 Kyofu a formidable fighter aircraft.

With all the tests satisfactorily completed, the Navy ordered the production of the aircraft, but production was extremely slow and by December 1943 the output was just fifteen aircraft per month. Because the N1K1 was deemed to be an offensive ground support aircraft and Japan was now being forced to defend itself, the decision was taken to cease production in favour of a land-based version, and a total of only eighty-nine were built.

In December 1941, the Kawanishi company, who were now concentrating on the production of floatplanes and flying boats, were now asked to produce a single-seat land-based interceptor version of their N1K1 *Kyofu* floatplane. The performance of the floatplane warranted an attempt to produce a land-based version, as very few modifications were required. The ventral float and the two outrigger floats were removed and replaced with a long-legged retractable undercarriage, which retracted into the 39 ft 4½ in. wings that were mounted in the mid-fuselage position. The 1,460-hp, 14-cylinder Kasei engine in the floatplane was replaced with the 1,820-hp, 18-cylinder Nakajima Homare 11 radial engine, and the three-bladed propeller was replaced with a four-bladed version that had a 10 ft 9½ in. (3.3 m) diameter.

Within twelve months the first prototype took to the air, with only a few minor flight problems to contend with. This was just seven months after the maiden flight of the N1K1 floatplane, a quite remarkable achievement. There was one major problem: because of the large-diameter propeller, the undercarriage had to have extended legs. This in turn caused the pilot's visibility while taxiing to be very limited. Four prototypes were built and one was handed over to the Navy for evaluation.

At first the Navy test pilots and the engineers were very critical, pointing out that although it was very manoeuvrable, its top speed was not what the company had claimed and its design left a lot to be desired. Reluctantly, they had to admit that it was faster than anything that they had at the time and it also had a longer range. The Pacific War had started by this time and the Navy needed a fighter that was capable of standing up to the US Navy's Grumman Hellcats and Vought Corsairs.

Designated the Kawanishi N1K1-J Shiden, the aircraft was put into production and armed with two 20 mm cannons installed in the wings and two under-wing 20 mm cannons mounted in gondolas, together with the two fuselage-mounted 7.7 mm machine guns. Seventy N1K1-J aircraft were produced by the end of 1943 and almost immediately saw action over the Philippines. The aircraft was a superb aircraft in the air, but on the ground difficulties with the landing gear still plagued it. For experienced pilots there was no problem, but for the novice pilots it could be a nightmare, and as the war progressed Japan had to produce more and more pilots with limited training.

Given the Allied name 'George', the production and development of the aircraft was stepped up and a new model appeared, the N1K1-Ja, which had four 20 mm cannons mounted in the wings and no fuselage-mounted machine guns. This was followed by the N1K1-Jb model, which had four improved 20 mm cannons mounted in the wings and bomb racks fitted under each wing capable of carrying 551 lb (250 kg) bombs.

A new version appeared in 1943, the N1K2-J. This was developed to eliminate the long-legged undercarriage that was causing so many problems for the novice pilots, of which there were now a large number. The wings were moved from their original position at mid-fuselage to the lower fuselage, thus making the undercarriage the conventional length. The fuselage and tail surfaces were re-designed, leaving only the wings and armament of the original aircraft.

Teething problems with the engines caused delays in production, but in June 1944 the Kawanishi N1K2-J Shiden Kai was handed over to the Navy. This aircraft was one of the most outstanding Japanese fighter aircraft in the Pacific War, and proved itself to be the equal of any of the Allied fighters with which it came in contact. Its only problem was that this late in the war it did not have the pilots to show its capabilities. A total of 1,435 N1K1-J and N1K2-Js were built between 1942 and 1944.

ANTI-SUBMARINE PATROL

In 1942, with the increased number of American submarines operating in the Pacific Ocean, the Japanese Navy saw the need for a purpose-built anti-submarine patrol aircraft and approached the Watanabe company. The specifications called for a three-seat aircraft that had long-range capabilities and could operate at reasonably low speeds. It also had to have the attributes of a dive-bomber and be able to negotiate safely over vast expanses of sea.

The proposal put forward by the Watanabe company was for a design of a twin-engined, low-wing aircraft of a wood and steel construction, which was based on the design of the Junkers Ju-88, with a cockpit that gave all the crew members a good all-round view. It was powered by two 610-hp Hitachi GK2C Amakaze 31, 9-cylinder, air-cooled radial engines, which were fitted into specially designed wings that had a constant taper on the leading and trailing edges. In addition, the aircraft was fitted with a Type 3 radar system that was connected to a magnetic anomaly detection gear. The only defensive armament carried was a single, flexible rear-firing 7.7 mm machine gun. Offensive armament was two 551 lb (250 kg) depth charges and provision was made for the installation of two 20 mm cannons.

The prototype was passed to the Navy for test and evaluation and was found to be an extremely easy and pleasant aircraft to fly, with good handling qualities and manoeuvrability. In the spring of 1944, the Navy gave the aircraft official designation of Navy Patrol Aircraft Tokai ('Eastern Sea') Model 11 (Q1W) and ordered the aircraft into production. Given the code name 'Lorna' by the Allies, the Q1W operated from bases in Japan, China and Formosa, escorting convoys bringing urgent supplies from the Dutch East Indies.

The aircraft's slow speed, very limited defensive armament and almost non-existent armour made them sitting targets for the heavily-armed Allied fighters that were now operating from aircraft carriers in the Pacific. Between 1943 and 1945, a total of 153 Q1W aircraft were built.

THE *STORCH* REIMAGINED

One of the lesser-known of Japan's aircraft manufacturers, the Tachikawa company, was ordered to design and develop a fast, two-seat, single-engined monoplane with a short take-off and landing capability enabling it to operate from small rough strips just behind the front line. It was to be used for reconnaissance purposes and had to have good downward visibility for the pilot and observer. It also had to be extremely manoeuvrable at low altitude and have provision for photographic and radio equipment, together with bomb racks mounted beneath the wings.

Powered by a 450-hp Hitachi Ha-13 9-cylinder radial engine, the Tachikawa Ki-36, as it was known, was a low-wing monoplane with a lightweight airframe, a large wing area and large-sized elevators and rudder together with a fixed spatted undercarriage. These factors combined gave the aircraft a very sensitive control factor, which when flown at low speeds and low altitude was very necessary.

The first prototype was sent for test and evaluation, but a problem was discovered when the Ki-36 suffered from wing-tip stall. This was easily resolved by fitting leading edge slots on the second prototype. Both prototypes were fitted with fixed forward-firing 7.7 mm machine guns, with an additional 7.7 mm machine gun mounted inside the engine cowling and slightly offset to the starboard. The observer's position in the rear cockpit was also fitted with a rearward-firing, flexible 7.7 mm machine gun. Bomb racks mounted beneath the wings allowed for a variety of small bombs and anti-personnel bombs to be carried.

Given the designation Army Type 98 (Ki-36), its first encounter with the enemy was in the second Sino-Japanese conflict, where it performed well. Because of its easy handling characteristics, it lent itself to use as an advanced trainer. The Tachikawa company took up the project and removed all unnecessary equipment, including the spats that covered the undercarriage. Designated the Army Type 99-Advanced Trainer (Ki-55), it was sent to Army Flying Schools and, towards the end of the Second World War, used on suicide missions. A total of 1,334 Ki-36 reconnaissance aircraft were built and 1,389 Ki-55 advanced trainers built between 1938 and 1942.

With the success of the Ki-36, information filtered back about a remarkable reconnaissance aircraft built by the Fieseler company in Germany, the Fieseler Fi-156 *Storch*. The Army realised the immense potential in such an aircraft and approached a small aviation company, Kokusai Koku Kogyo, to build a similar aircraft. The designer Kozo Masuhara set to work and produced the Kokusai Ki-76, which although looking very similar to the German Fieseler Fi 156 Storch, produced a much better performance except for the landing distance, which considerably longer. Powered by a 310-hp Hitachi Ha-42 9-cylinder air-cooled radial engine, the Ki-76 was fitted with Fowler flaps, as opposed to the slotted flaps of the Storch, which were synchronised with the variable-incidence horizontal tail surfaces. This gave the aircraft a higher lift coefficient.

During the initial flight trials, it was realised that the aircraft was so easy to fly that pilots with very limited flying experience could handle the aircraft with consummate ease. The number built is not known, but the Ki-76 was also used for Anti-submarine Warfare (ASW) missions along the coastline of Japan towards the end of the war.

The JNAF also took note of the Ki-76 and its relatively short take-off and landing capability, and ordered two of the aircraft. Both the aircraft were fitted with an arrester hook and flown aboard the aircraft carrier *Akitsu Maru*. The idea was to use them as a flying observation platform, but because of the lightness of the aircraft they became extremely difficult to handle during deck landings and they were quickly phased out.

When a new competition was instigated in 1936 by the JAAF, the Type PE was entered as the Army Type 97 Ki-27. Designed by Yasushi Koyama, the Type Ki-27 was similar in design to the Ki-11 but had an enclosed cockpit. This was a low-wing, all-metal fighter aircraft, the first of its type to be built. Two prototypes were built and submitted to the Army Aerotechnical Research Institute (RKG – Rikugun Kokugijutsu Kenkyujo) and pitted against the Mitsubishi and Kawasaki entries. After extensive testing the Ki-27 was selected and the aircraft put into production. The first ten went through further testing, resulting in modifications being made by increasing the wingspan and modifying the cockpit canopy.

The aircraft was armed with a pair of synchronised 7.7 mm machine guns mounted in the forward decking of the upper fuselage. The aircraft was immediately put into mass production as the war against China rapidly escalated. On 4 May 1939, Japanese and Russian forces clashed on the Manchuko/Outer Mongolian border, and the Army sent five squadrons, the 1st, 11th, 24th, 59th and 64th, all equipped with the Nakajima Army Type 97 Ki-27 fighter. A total of over 200 of these aircraft were in action during this campaign.

THE ARMY IMPROVES PERFORMANCE

The hierarchy of the Army was obsessed with the idea that to be successful, a fighter aircraft had to be very manoeuvrable and so dismissed many sound aircraft on the grounds that they did not meet the levels required. This was never more apparent than in the competition won by the Nakajima Army Type 97 Ki-27. One of the other competitors was the Kawasaki Ki-28, which was a high-speed fighter with a high wing loading that had been designed specifically for hit-and-run tactics. It was never developed because of its lack of manoeuvrability, but there is no doubt that it would have been a valuable tool in the armoury of the Japanese Army.

The success of the Ki-27 during the Sino-Japanese War was one of the factors that led the Army to concentrate on developing more manoeuvrable aircraft, which eventually allowed the performance of its fighter aircraft to fall way behind those of other countries. The Ki-27 remained in production until the beginning of the Pacific War, and a total of 3,386 of these fighters were built.

The increasing demand for fighters and bombers overshadowed the need for long-range reconnaissance aircraft. The Koku Hombu recognised this, and although they had the Mitsubishi Ki-15, they also recognised its limitations and issued Mitsubishi a requirement for a successor. The specification called for a long-range photographic and reconnaissance aircraft with a minimum six-hour endurance at a cruising speed of 250 mph (400 km/hr) and a height of between 13,125 ft (4,000 m) and 19,685 ft (6,000 m). The aircraft had to have a top speed of 373 mph (600 km/hr) and a flexible rear-firing 7.7 mm machine gun.

Mitsubishi's designer Tomio Kubo began preliminary design studies, calling upon the Aeronautical Research Institute of the University of Tokyo to help streamline the aircraft. The aircraft was to be powered by two Mitsubishi Ha-126 14-cylinder radials and the institute came up with close-fitting cowls for the two engines that not only reduced the drag but also improved the pilot's sideways visibility and incorporated the retractable landing gear. The two-man crew (pilot and radio operator/gunner) were seated in two cockpits, which were separated by the fuel tank that was mounted in the fuselage.

In November 1939, the prototype Ki-46, as it was known, had been sent to the Army for test and evaluation. All the tests were completed satisfactorily, but the aircraft failed to meet its proposed design speed. This, however, did not prove to be a stumbling block, because it was still faster than the latest fighter that was about to be delivered to the Army. It was also faster than the Mitsubishi A6M2 that the Navy had just taken delivery of, and because of the intense rivalry, that was very pleasing to the Army.

Further tests under hot and humid conditions found that vapour locks occurred frequently. Using a higher-octane fuel, from 87-octane to 92-octane, and re-routing the fuel lines around the engine, resolved the problem. There were a number of other minor issues, but one major problem that manifested itself affected the undercarriage. Because of the aircraft's high sink rate the undercarriage often collapsed on landing, and despite the fitting of a much stronger auxiliary rear strut, the aircraft continued to have a weak undercarriage throughout its operational life.

Despite this problem, the Mitsubishi Ki-46–II, as it was designated, attracted attention from the IJNAF, which managed to persuade the Army to let it have a small number. Even the German Luftwaffe attempted to get a licence to manufacture the aircraft under the Japanese-German Technical Exchange Programme, but they were unsuccessful.

With the outbreak of the Second World War, the Ki-46 carried out reconnaissance missions as far as the Bay of Bengal and over Northern Australia. However, when the USAAF deployed a squadron of P-38F Lightning fighters into the Pacific Theatre, and the Australians were equipped with Spitfire Mk Vs, the Ki-46 was no match for either of these Allied aircraft and their losses started to mount. Aware of this, the Koku Hombu instructed Mitsubishi to install the new 1,500-hp Ha-112-II fuel-injected engine. This meant that an additional fuel tank had to be installed in front of the pilot and the engine nacelles had to be enlarged. The pilot's canopy was re-designed so as to sweep forward, encompassing the step between the nose and the fuselage. The landing gear was strengthened because of the increased weight, but was still the cause of some concern with the crews. Designated the Ki-46-III, the improvement in performance was to be the cause of some concern for the Allies, as it was now able to climb faster and higher and it was only in the later stages of the war that Allied fighters under radar control could catch it.

A number of different production models were built, including a pilot training version. A navigator training model and a high-altitude interceptor fighter version appeared towards the end of the war. The Mitsubishi Ki-46 was one of the most successful reconnaissance aircraft that Japan built, and a total of 1,742 of the aircraft were built between 1939 and 1944.

In May 1937, the Army had initiated a six-year expansion programme for its air units and set up a number of flight schools. The Army Juvenile Flying School was set up to take boys as young as fifteen to be trained as pilots. In 1939, four more schools were set up: Gunnery-Communication, Basic Flight Training, Navigation and Light Bomber, bringing the number of flight training schools to eight. This expansion programme heralded the start of a massive military build-up, and by the end of 1940, the Army had twenty reconnaissance squadrons, thirty-six fighter squadrons, twenty-eight light bomber squadrons, twenty-two heavy bomber squadrons and one super-heavy bomber squadron, a total of 107 operational squadrons.

The Navy had not been idle; it had thirty-one operational squadrons, three (386 aircraft) land-based, twenty-three squadrons at training units (446 aircraft) and fifty-three operational units aboard ships (1,915 aircraft).

THE ULTIMATE JAPANESE FIGHTER?

In December 1937, the Army had approached the Nakajima company with a request for a new fighter to replace the Army Type 97 Ki-27. Led by designer Hideo Itokawa, Nakajima set about producing one of the best fighter aircraft Japan ever produced. Within one year, the first prototype was rolled out of the factory in complete secrecy. The initial tests were extremely successful, and two more prototypes emerged at the beginning of 1939. Powered by a 925-hp Nakajima Ha-35 radial engine, the Army Type I Fighter Model I Hayabusa (Ki-43) ('Peregrine Falcon' – 'Oscar' to the Allies) was sent to the JAAF for test and evaluation. One of the features of the Ki-43 was that it had a fully retractable undercarriage and a telescopic gunsight protruding through the windshield.

The evaluation tests went well, but once again the fixation the Army had regarding manoeuvrability raised its head. The pilots who carried out the tests complained that the aircraft lacked the manoeuvrability of some of the earlier models, because they still believed that the most manoeuvrable aircraft would win in air combat. Nakajima set about improving the model and a further ten prototypes were built, all with only minor modifications, like an all-round vision canopy and a larger engine with a two-speed supercharger. Some of the later prototypes were fitted with 'butterfly' combat flaps, which, when extended in a combat situation, increased control sensitivity, enabling the pilot to carry out a much tighter turning circle. It was the latter modification that convinced the Army pilots this aircraft should be put into immediate production.

There was one major flaw in the design – a lack of any armoured protection for the pilot and the fuel tanks. This was to cause the demise of many Army pilots in combat situations.

The aircraft did not enter production until November 1942, and was given the designation Nakajima Army Type1 Fighter Model 2A (Ki-43-11A). There were a number of minor modifications made during the production years, including the fitting of various engines. A total of 5,919 of the aircraft were built, 3,241 by the Nakajima company, 2,629 by the Tachikawa Hikoki company and forty-nine by the Dai-Ichi Rikugun Kokusho company.

At the same time as the Ki-43 was being developed, the Nakajima company was working on a high-speed interceptor fighter, the Ki-44. It was put forward for test and evaluation by the Army's test pilots, and once again was rejected on the grounds of manoeuvrability. Despite this setback, the Nakajima company continued to develop the aircraft under the guidance of their project engineer T. Koyama.

Powered by the 1250-hp Nakajima Ha-41 engine, the Ki-44 could reach an altitude of 16,450 ft (5,000 m) in 4 mins 17 secs, and was the fastest climbing Japanese fighter. Once again it was rejected by the Army, which later realised the need for a high-speed interceptor and ordered the production of the aircraft. Designated the Army Type 2 Fighter Model (Ki-44) Shoki, it too had a retractable undercarriage and was fitted with 'butterfly' combat flaps. It also carried a single drop-tank mounted beneath the fuselage centreline. Once in production a number of minor modifications were introduced, including the fitting of a 1,540-hp Nakajima Ha109 engine, two synchronised 7.7 mm machine guns and two wing-mounted 12.7 mm machine guns. Later in the war, when B-29 bombers started raids against the Japanese mainland, a number of the Ki-44s were fitted with wing-mounted 40 mm Ho-31 cannons.

A NEW GERMAN ENGINE

Between November 1940 and December 1944, a total of 1,225 Ki-44s were built and were chosen over the Messerschmitt BF109E, a decision never regretted. The one thing that the Army realised was that it did not have high-performance, liquid-cooled engines, and so it told Kawasaki to enter into negotiations with the Daimler-Benz company in Germany for the licence to build the DB 601A engine in Japan. The Aichi company was already in negotiations with Daimler-Benz over the same engine for the Navy.

One of the first aircraft to emerge powered by the DB 601A engine was from Kawasaki in February 1939. It was a high-speed fighter that bore a strong resemblance to a European-designed aircraft. Known as the Ki-60, the prototype was sent to be tested by service pilots, but was rejected on the grounds that it wasn't as fast as stated by Kawasaki. It had an exceptionally fast landing speed because of its high wing loading, but was nowhere near as manoeuvrable as other fighters.

Using the information gained from the tests of the Ki-60, Kawasaki produced the Ki-61. This was the first fighter aircraft to incorporate the Ha-40 12-cylinder liquid-cooled engine, armour protection for the pilot and the fuel tanks, and a retractable undercarriage and tailwheel. The prototype was sent for test and evaluation. It impressed the test pilots with its high diving speed and was accepted immediately and put into production. Given the designation Kawasaki Army Type 3 Fighter Model 1A (Ki-61), it was armed with two fuselage-mounted 12.7 mm machine guns and wing-mounted 7.7 mm machine guns. Later models were fitted with 20 mm cannons in both the fuselage and the wings. Given the name Hien ('Swallow') by the Japanese, it was codenamed 'Tony' by the Allies.

The first eleven Ki-61s off the production line were considered prototypes by the Army and subjected to intense handling and performance tests before being sent to pilot conversion and training units at Ota. At the beginning of 1942, the 68th and 78th Sentai (Corps) were sent to the north coast of New Guinea and New Britain and were equipped with the new Ki-61 fighters. Within days of going into action against American and Australian fighters it proved to be a worthy adversary and quickly gained the respect of the enemy.

Problems were discovered with the handling of the aircraft because of the hot, humid conditions in New Guinea and New Britain, so the ease of maintenance had to be improved. This information was passed back to Kawasaki, who decided to strengthen and simplify the structure of the aircraft in an effort to ease the maintenance problem. Between 1942 and 1945, 3,078 Ki-61 aircraft were built.

Back in 1938 the JAAF had obtained authorisation from the Koku Hombu to develop a twin-engined light bomber. This had come about when, during the second Sino-Japanese war, the appearance of the Russian Tupolev SB-2 caused them major problems. It was almost immune to fighter interception and was as fast as the latest Nakajima Ki-27 single-engined fighter. This impressed the Japanese Air Wing commanders, and they asked for a similar aircraft to be built with a maximum speed of 298 mph (480 km/hr) at 9,845 ft (3,000 m), a cruising speed of 217 mph (350 km/hr) at the same height, a bomb load of 882 lb (400 kg) and armament of three flexible 7.7 mm machine guns. It also had to have the ability to operate under extreme winter weather conditions.

The appearance of the Russian Tupolev SB-2 during the second Sino-Japanese conflict caused the Japanese to look hard at their own aircraft. The SB-2 was superior to anything that they had at the time, and so impressed with the aircraft were the Japanese that they instructed the Kawasaki company to design and build a twin-engine light bomber that could compete with the Russian aircraft.

The specifications laid down were quite onerous: the bomber had to have a maximum speed of 298 mph, have a climb rate to 1,640 feet a minute, be able to carry a bomb load of 882 lbs, carry a crew of four, have armament of between three and four flexible machine guns and be operational in the extreme cold weather conditions of the Manchukuo-Siberian border.

Using the information gained in the design of the Ki-45, the design team got to work. The first of four prototypes appeared in July 1939 and, after extensive testing by service test pilots, was accepted. The performance met all the requirements laid down by the Koku Hombu and the handling characteristics and manoeuvrability of the aircraft drew admiration from all those who flew it. There was one problem that was the cause of some concern, and that was that all the prototypes suffered from severe tail flutter. The first five prototypes were used to find the cause of the problem and by strengthening the rear fuselage section and raising the horizontal tail surfaces 13¾ in. (40 cm), the problem was finally solved.

With the problem solved, the Ki-48, as it was designated, was put into production, the first one coming off the production line in July 1940. Given its official designation of Army Type 99 Twin-Engined Light Bomber 1A (Ki-48-1a), within weeks it had been assigned to the 45th Sentai in Northern China. Its baptism of fire was almost non-existent, as the Chinese Air Force had virtually no aircraft, so the Ki-48-1a was eased gently into the war scenario. This was to give the crews a false sense of security because when they did encounter the odd Chinese fighter, they were able to deal with it easily. Preparations were made to use the Ki-48 as a night bomber, with increased armament and a few other minor changes making it the Ki-48-1b.

The aircraft production increased with the onset of the war, and by 1942 a total of 557 Ki-48-1a and 1bs had been built and assigned to Sentais in Malaya and Burma. This was a rude awakening for the crews of the Ki-48, because up to this point in time they had faced no opposition, but now they were facing Allied fighters that were faster, more manoeuvrable and more heavily armed. The weaknesses in the Ki-48 were exploited very quickly. It was too slow, its defensive armament was woefully inadequate, the bomb load was insufficient and the crew and fuel tanks lacked protection.

The losses started to mount and the Ki-48 was reduced to night operations. Kawasaki, in the meantime, was developing an improved model with better armament and more protection for the crew and fuel tanks. The engines were replaced with two Nakajima Ha-115 14-cylinder, air-cooled radials with two-stage blowers. The aircraft was put into production as the Army Type 99 Twin-engined Light Bomber Model 2A (Ki-48-11a). The fuselage had also been strengthened and it had been fitted with retractable dive brakes, enabling it to be used as a dive-bomber. Even with all these modifications it was still too slow to avoid interception, and soon became easy prey to the Allied fighters. Its defensive armament was increased, but by the end of 1944 it was obsolete and 1,408 had been built. What aircraft remained were relegated to be used as suicide bombers, known as Army Type 99 Special Attack Planes, and for a variety of other test programmes, including the fitting of turbojets. This was done by removing the bomb bay doors and placing the turbojet under the fuselage.

JAPAN'S PROBLEMS

During the period the Ki-48 was being built, the number of new models coming out of the manufacturers was increasing, and the Koku Hombu and its Navy equivalent Kaigun Koku Hombu (Naval Bureau of Aeronautics), appeared to be rejecting them almost out of hand. One example of this was the Kawasaki single-seater fighter Ki-64 ordered by the Army, which had two Ha-40 engines mounted in tandem, causing almost insurmountable problems. So much time was wasted in trying to get this project to work that other projects went by the wayside. It appears that the manufacturers were at the whim of the Army and Navy, and if either of them got a project into their heads they insisted that the manufacturers not be diverted until all the problems were solved. Had the manufacturers been able to concentrate on proven aircraft designs and engines, then the respective airpowers would have been strengthened considerably.

The reason for Japan's unprovoked attack on the United States of America on the morning of 7 December 1941 has always been the subject of debate. The truth of the matter is that it essentially came down to the one commodity that even today rules the way that the major countries think and behave – oil, the lifeblood the world has come to rely on.

At the beginning of 1941, the war in China was rapidly using up the oil stocks in Japan. The total reserve stock of aviation fuel stood at around 300 million US gallons, the stock of crude petroleum at 1,130 million US gallons. It was estimated that by the end of 1943, oil stocks would be reduced by half unless the situation was resolved one way or another. The Americans were demanding a complete withdrawal of armed forces from China and the scrapping of the Tripartite Pact that Japan had signed with Germany and Italy, both of whom were at war in Europe. Failure to do this would result in the cutting off of oil supplies to Japan, among other sanctions.

The rumblings of war in the Pacific would in essence mean that there would be no oil coming from America, which was Japan's main supplier, or any of her Allies – and that left Indonesia as the only country that could keep Japan supplied with oil. In the event of a war, Japan's oil stocks would only last for a year so it was decided that the Japanese military would have to carry out an attack and take over the oil fields in Indonesia.

At the same time as the attack on Pearl Harbor was being carried out, the Japanese Army invaded mainland Malaya under the cover of air attacks by squadrons based in Indochina. Now there was no going back, and all negotiating doors had been firmly closed.

A SECOND SUBMARINE-LAUNCHED BOMBER

On 15 May 1942, the Kaigun-Koku-Hombu gave the Aichi company the specifications for the aircraft that was to be called the Experimental 17 Shi Special Attack Bomber. The term 'Shi' is a shortened version of the term *Shisauku Seizo* ('Trial Manufacture') and the term 'Special Attack' was invariably associated with kamikaze aircraft, such as the Navy's Ohka ('Cherry Blossom') rocket-propelled suicide aircraft, which could only be described as a bomb with wings. It was the intention of the Navy to use the Experimental 17 Shi Special attack aircraft for the following purpose: this aircraft was to be the first purpose-built submarine-based aircraft that had a strike mission as its primary role. Although the design of the aircraft was similar to that of the carrier-based dive-bomber, the Yokosuka D4Y Suisei, it became necessary to redesign it so that it would fit into an 11 ft 6 in. diameter hangar on a submarine. There was a six-inch clearance at the tip of each propeller when the aircraft was inside the hangar; this did not leave a lot of room for mishandling. The name given to this aircraft was the Aichi M6A1 Navy Special Attack Bomber Sieran ('Mountain Haze').

The engine selection was both critical and limited. Three different engines were considered: the Mitsubishi Mk 8 of 1,300 hp, the Nakajima NK 9B Homare of 1,800 hp, and the Aichi Atsuta 32 of 1,400 hp. The first two were radial cooled engines and were considered the most reliable, but their relative sizes caused a problem. Because of the limited space between the bottom of the fuselage and the hangar deck, it was impossible to hang a torpedo or bomb from the fuselage, and it was necessary to have the armament in place when the aircraft was stowed. Also to be taken into consideration was the fact that when the submarine surfaced to launch its aircraft, the submarine was at its most vulnerable and time was of the essence. It was therefore much safer to arm the aircraft while it was in its hangar and the submarine was submerged.

Aichi made a full-scale mock-up of the aircraft out of wood, which enabled it to work out any problems with the intricate design. The main design feature that had to be perfected was the folding back of the wings against the fuselage. The method devised by Aichi had the wings pivoting on the main spar where it joined the fuselage, and then by rotating downwards, the wing could lay flat against the fuselage, much like the Grumman Avenger. The top part of the rudder folded down to give it hangar clearance, and the two pontoon floats were removed and housed in their own container below the level of the deck. When launching, the whole of the aircraft was lowered almost to track level by collapsible, compressed air-activated struts on the heavy catapult car. The cockpit was designed for a crew of two, and they were housed in a veritable greenhouse, giving them all-round visibility. The armament carried, besides the bombs or torpedoes, was a 13 mm Type 2 machine-gun mounted in the rear cockpit. With the detachable floats, folding wings and tail surfaces, it was estimated that four fully trained men could prepare the Sieran for launch in less than seven minutes.

The 4,500-ton I-400 Class submarines that were used to carry the Aichi M6A were capable of carrying three of the aircraft. They had a cruising radius of 41,575 nautical miles at a speed of 14 knots. Five of the submarines were built but none ever saw action.

As the war progressed the need for pilots increased. The number of experienced Japanese fighter pilots was diminishing by the day, leaving the new, inexperienced pilots to carry the fight to the Allies. The need for training these new pilots prompted the Mitsubishi company to put forward a proposal that some of the existing single-seat, front-line fighters be converted into trainers, thus enabling them to carry out a dual role. The JNAF took up the suggestion, and ordered that the trainer version be put into the existing 17-Shi A6M2 production lines.

The main airframe was retained, but a number of modifications were made, including the installation of a two-seat cockpit fitted with dual controls, the mounting of a small horizontal fin on either side of the rear fuselage section, and the removing of the two wing-mounted 20 mm cannons. The cannons could easily be refitted, as provision was made for them. The first of the new trainer fighters, the Mitsubishi A6M2-K, rolled off the production line in November 1943. Another version, the A6M5-K, was produced by the Hitachi company, but they were seven experimental models that had shorter wings but were identical in every other way. A total of 515 A6M2-Ks were built between 1943 and 1945.

A NEW BOMBER AND NIGHT-FIGHTER

While the Army was introducing its latest fast twin-engined bomber, the Navy was looking towards developing its own. The tide of war was rapidly changing for the Japanese; no longer were they the aggressors, they now were fighting desperately for their survival. Control of the air was rapidly becoming the domain of the Allies and American bombers were beginning to pound the Japanese homeland. In an effort to help stem the Allied tide, the Yokosuka Arsenal produced a Naval version of the Army's Ki-67 Hiryu, the Yokosuka P1Y1 Ginga ('Milky Way'). This was a fast, twin–engined medium bomber of an all-metal construction, powered by two 1,820-hp Nakajima Homare 18-cylinder air-cooled radial engines.

The requirements were for a fast bomber that was capable of carrying out low-level attacks in addition to carrying out torpedo and dive-bombing missions. Comparable aircraft were the Junkers Ju-88, the North American B-25 Mitchell and the Martin B-26 Marauder. The wings were mounted mid-fuselage and contained eight protected and six unprotected fuel tanks containing a total of 1,218 gallons (5,535 litres) of fuel. In addition to this, provision was made for two drop tanks containing forty-eight gallons (220 litres) each. But, typical of Japanese design, the amount of armour in the aircraft was restricted to a small 20 mm plate behind the pilot's head. Armament consisted of one flexible 7.7 mm machine gun mounted in the nose and one in the rear cockpit. Provision was made for two 1,102 lb (500 kg) bombs or a single 1,746 lb (800 kg) torpedo mounted semi-internally under the fuselage.

In August 1943 the prototype made its first flight, and received very favourable reports from the test pilots. It was then sent to the Navy for test and evaluation of the aircraft's requirements, all of which exceeded expectations. There was no such response from the ground crews, who found the maintenance of the aircraft extremely troublesome, especially the hydraulic system and the power plants. These maintenance problems continued to plague the production of the P1Y1, causing a number of modifications to be made. A number of changes were also made, including the fitting of a flat bulletproof windscreen in front of the pilot, in place of the curved glass one. Other changes included the introduction of flush rivets on the fuselage and wings, giving the aircraft a slightly more streamlined look, the fitting of a fixed tailwheel instead of a retractable one, and a new exhaust system and engine cowlings. The production models were fitted with the new 1,825-hp Nakajima Homare 12, 18-cyinder, air-cooled radial engines.

After further extensive testing the Navy accepted the aircraft, but reliability problems with the Homare 12 engine kept the aircraft from the front line until the beginning of 1945. Despite its shortcomings, the Navy was impressed with its manoeuvrability and its maximum speed of 340 mph (295 knts). The Americans had just started night bombing missions on Japan, and the Navy decided that the P1Y1 would be the ideal aircraft to be turned into a night-fighter interceptor. Instructions were given for a number of the P1Y1s to be converted into night-fighters and the Kawanishi company was ordered to carry out the requirements.

The machine gun in the nose of the aircraft was removed, and a 20 mm Type 99 cannon was mounted on either side of the fuselage and angled obliquely to fire upwards and forwards. The engines were replaced with two 1,850-hp Mitsubishi Kasei 25a, 14-cylinder radials. Designated the P1Y1-S, it went into production as the Navy Night Fighter Kyokko ('Aurora') and ninety-six of the aircraft were built. It was not a success. Its performance at altitude against the heavily-armed bombers and their fighter escorts proved to be disastrous, so the angled cannons were removed and the aircraft reverted back to their original role.

Towards the end of the war, a number of modifications were made in an attempt to produce something that would help stem the tide of relentless bombing, but the war ended before any of them could be tried. Between 1943 and 1945, a total of 1,098 P1Ys were built.

When the USAAF started to make daylight-bombing raids on mainland Japan, the Japanese Army found itself in desperate need of a high-altitude fighter capable of climbing to 30,000 feet and engaging the B-29 Superfortress bombers. Time was rapidly running out for the Army as the tide of war was turning against them. One of the stumbling blocks was finding an engine of small enough diameter to put into a fighter that had the capability to fly and fight at the height of 30,000 ft.

The only one available was the 1,500-hp Mitsubishi Ha-112-II 14-cylinder radial engine. The airframe chosen for this aircraft was the Ki-61-II, but the width of the fuselage was too small for the engine. Fortunately the Japanese engineers had access to a Focke-Wulf FW 190A, and were able to see how German engineers had managed to fit a large radial engine into a slim fuselage. The result was the Kawasaki Army Type 5 Model 1B (Ki-100).

The prototype was sent for test and evaluation and performed better than expected. It had a lower speed than the Ki-61-II, but its manoeuvrability and handling was markedly improved, which more than made up for it. Production of the Ki-61 was halted, and 271 of the airframes modified to become the Army Type 5. The first of the production models were quickly off the assembly lines and sent to the 5th, 17th, 18th, 59th, 111th and 244th Sentais in Japan. They proved to be more than capable of intercepting the high altitude B-29 bombers that roared over Japan during 1945 and dogfighting with the USN Grumman F6F Hellcats that operated from the aircraft carriers just off the coast of Japan. The Ki-100 came as a complete surprise to the Allies, and the Japanese Army regarded it as the most reliable operational fighter they had ever had. The fact that it was a most forgiving aircraft to fly enabled young pilots who had less than 100 hours of flight training time the opportunity to join operational units.

However, the Allied advance and the continuous bombing of Japanese manufacturing plants forced the production of the aircraft to cease. There were one or two modifications made to improve the performance of the aircraft, but none of them came to anything and by this time the war was all but over. A total of 396 Ki-100s were built in 1945, including 271 Ki-61-II airframes.

DEVELOPMENT OF A ROCKET PLANE

One of the most spectacular aircraft to appear in the 1940s was the Mitsubishi J8M Shusui ('Sword Stroke'). The need for such an aircraft came about because of the continuing high-altitude bombing by the Boeing B-29 Superfortress. The Japanese still had no aircraft capable of climbing to the B-29's bombing height, but fortunately the Japanese military attachés in Germany had become aware of a new rocket fighter that was capable of climbing to these heights, the Messerschmitt Me 163B. They obtained the manufacturing rights to build the aircraft and the Walter HWK 109-509 rocket engine.

In July 1944, two Japanese submarines were tasked to bring one of the engines and detailed drawings and technical data back to Japan. One of the submarines was tracked and sunk en route, but the other managed to make it back. The Mitsubishi company was tasked with designing and producing the aircraft and its engine and, in an extremely rare moment, the Army and the Navy became joint venturers in the project. The situation for Japan at the time was dire, and it needed everyone to set aside their differences and pull together, albeit in the case of the Army and Navy, it was a little too late.

Designated the J8M1 by the Navy and the Ki-200 by the Army, the mock-up appeared in September 1944 and three weeks later it was approved by both services and production started. The first of the prototypes arrived in December 1944, they and were glider versions. This was to provide data on the handling characteristics of the tailless aircraft, and to train pilots.

Lt-Com. Toyohio Inuzuka, the project pilot for the programme, piloted the first flight. The flight was a success and two more prototypes were ordered; one was sent to the Army's test facility at Tachikawa, and the other to the Navy's test facility at Nagoya.

After a series of glide tests, the first prototype with an engine installed was tested on 7 July 1945. The pilot, Lt-Com. Toyohio Inuzuka, put the aircraft into a steep climb just after take-off, but the engine suddenly failed and the J8M1 went out of control and crashed, killing the pilot. This set the programme back, and investigations to find out the cause of the crash ended abruptly as the war ended.

The Army ordered sixty of the aircraft, known as Ku-13, as heavy training gliders, and three light gliders designated MXY8. They were all built by various companies; Mitsubishi built just seven of the heavy models.

Even at the end of the first year of the war, the Japanese Army still regarded its air force as playing a secondary role. It still believed that their ground forces were the primary force and everything else was subservient. It was to take a further two years before the

Japanese Army was willing to accept that the power in the air was the key to winning the war, but by this time air power had shifted in favour of the Allies. Japanese supply lines were being cut dramatically and troops on the islands in the Pacific were not only running out of ammunition and reinforcements, but also food and medical supplies.

The Allies had almost gained complete control of the air in the Pacific, despite the frantic attempts by the Army to reinforce its depleted squadrons. In 1942, the Japanese Navy was decimated by the American Sixth Fleet in the Battle of Midway. During the engagement, it lost almost all of its major aircraft carriers, together with over 300 pilots and their aircraft. The Japanese Army found itself alone and retreating rapidly.

The Japanese Navy never really recovered from Midway and when, in June 1944, a large American task force descended upon the Mariana Islands to cover the amphibious landings of Saipan, Guam, Tinian and Iwo Jima, the Japanese Navy attempted to counterattack with just a token carrier force. In the Battle of the Philippine Sea, although they still retained a number of newer types of aircraft, the JNAF pilots were outclassed by the now battle-hardened American pilots and were decimated by what became known as the 'Marianas Turkey Shoot'. For the second time in a major engagement they lost over 300 pilots and their aircraft.

This catastrophic defeat caused the removal of the Japanese cabinet, headed by Hideki Tojo, and all the Japanese could do now was to brace themselves for an all-out onslaught.

SPECIFICATIONS

Nakajima Navy Type 15 Reconnaissance Seaplane (E2N1)

Wing Span:	44 ft 4¼ in. (13.5 m)
Length:	31 ft 4½ in. (9.56 m)
Height:	12 ft 1 in. (3.68 m)
Weight Empty:	3,106 lb (1,409 kg)
Weight Loaded:	4,299 lb (1,950 kg)
Max. Speed:	107 mph (93 kts)
Ceiling:	19,684 ft (6,000 m)
Endurance:	5 hours
Range:	651 sq. miles (566 nm)
Engine:	One 300-340-hp Mitsubishi Type Hi, 8-cylinder Vee, water-cooled
Armament:	One dorsal flexible 7.7mm machine gun

Yokosuka P1Y Navy Bomber *Ginga* ('Milky Way') (Frances)

Wing Span:	65 ft 7½ in. (20 m)
Length:	49 ft 2½ in. (15 m)
Height:	14 ft 1¼ in. (4.3 m)
Weight Empty:	16,017 lb (7,265 kg)
Weight Loaded:	23,149 lb (10,500 kg)
Max. Speed:	340 mph (295 knts)
Ceiling:	30,840 ft (9,400 m)
Endurance:	Not known
Range:	3,338 miles (2,900 nm)
Engine:	Two 1,820-hp Nakajima Homare 11, 18-cylinder, air-cooled radials
	Two 1,990-hp Nakajima Homare 12, 18-cylinder, air-cooled radials
	Two 1,825-hp Nakajima Homare 23, 18-cylinder, air-cooled radials
	Two 1,850-hp Mitsubishi Kasei 25a, 14-cylinder, air-cooled radials
Crew: Three	
Armament:	One flexible nose-mounted 13 mm machine gun.

Two 13 mm machine guns in dorsal turret
Bomb Load: One 2,205 lb (1,000 kg) or one 1,764 lb (800 kg) torpedo

KAWASAKI KI-32 ARMY TYPE 98 (MARY)

Wing Span:	49 ft 2 in. (15 m)
Length:	38 ft 1 in. (11.64 m)
Height:	9 ft 6½ in. (2.9 m)
Weight Empty:	5,179 lb (2,349 kg)
Weight Loaded:	7,802 lb (3,539 kg)
Max. Speed:	263 mph (423 km/h)
Ceiling:	29,265 ft (8,920 m)
Endurance:	3½ hours
Range:	1,218 miles (1,960 km)
Engine:	850-hp Kawasaki, 12-cylinder, water-cooled
Armament:	One forward-firing 7.7 mm machine gun mounted in engine cowling. One flexible rearward-firing 7.7 mm machine gun

NAKAJIMA KI-34 ARMY TYPE 97 TRANSPORT (THORA)

Wing Span:	65 ft 1 in. (19.90 m)
Length:	50 ft 2½ in. (15.3 m)
Height:	13 ft 7½ in. (4.15 m)
Weight Empty:	7,716 lb (3,500 kg)
Weight Loaded:	11,574 lb (5,250 kg)
Max. Speed:	224 mph (360 km/h)
Ceiling:	22,965 ft (7,000 m)
Endurance:	3½ hours
Range:	745 miles (1,200 km)
Engines:	Two 580-hp Nakajima Kotokubi 41, 9-cylinder, air-cooled radials. (AT-1) Two 780-hp Nakajima Kotokubi Ha-1b 9-cylinder, air-cooled radials (AT-2)
Armament:	None

MITSUBISHI NAVY TYPE 97 CARRIER ATTACK BOMBER B5M (MABEL)

Wing Span:	50 ft 2½ in. (15.3 m)
Length:	33 ft 10½ in. (10.3 m)
Height:	14 ft 2 in. (4.32 m)
Weight Empty:	4,643 lb (2,106 kg)
Weight Loaded:	8,819 lb (4,000 kg)
Max. Speed:	237 mph (205 knts)
Ceiling:	27,100 ft (8,260 m)
Endurance:	7 hours
Range:	1,460 miles (2,350 km)
Engine:	One 1,000-hp Mitsubishi Kinsei 43, 14-cylinder, air-cooled radial
Armament:	Two fixed forward-firing 7.7 mm machine guns mounted in wings and one flexible 7.7 mm machine gun in rear cockpit; 1,700 lb (771 kg) of bombs
Crew:	Three

NAKAJIMA B5N NAVY TYPE 97-1&3 CARRIER ATTACK BOMBER (KATE)

Wing Span:	50 ft 10½ in. (15.51 m)
Length:	33 ft 9½ in. (10.3 m)
Height:	12 ft 1½ in. (3.7 m)

Weight Empty: 4,643 lb (2,106 kg)
Weight Loaded: 8,157 lb (4,015 kg)
Max. Speed: 229 mph (368 km/h)
Ceiling: 24,280 ft (7,400 m)
Endurance: 7 hours
Range: 679 miles (1,092 km)
Engine: One 700-hp Nakajima Hikari 3, 9-cylinder, air-cooled radial (B5N1)
                 One 1,000-hp Nakajima NK-1b Sakae 11 14-cylinder, air-cooled radial (B5N2)
Armament: One flexible 7.7 mm machine gun
                 1,764 lb (800 kg) of bombs or one 1,764 lb (800 kg) torpedo

Nakajima Carrier Attack Bomber B6N1/2 Tenzan (Jill)

Wing Span: 48 ft 10½ in. (14.84 m)
Length: 34 ft 0½ in. (10.36 m)
Height: 12 ft 1½ in. (3.7 m)
Weight Empty: 7,105 lb (3,223 kg)
Weight Loaded: 11,464 lb (5,650 kg)
Max. Speed: 289 mph (465 km/h)
Ceiling: 28,380 ft (8,650 m)
Endurance: 7 hours
Range: 909 miles (1,462 km)
Engine: One 1,800-hp Nakajima NK7A Mamoura, 14-cylinder, air-cooled radial (B6N1)
                 One 1,850-hp Mitsubishi MK4T-C Kasei 25C 14-cylinder, air-cooled radial (B5N2)
Armament: One flexible 7.7 mm machine gun
                 1,764 lb (800 kg) of bombs or one 1,764 lb (800 kg) torpedo

Nakajima Navy Carrier Reconnaissance C6N1 Saiun (Myrt)

Wing Span: 41 ft ½ in. (12.5 m)
Length: 36 ft ½ in. (11 m)
Height: 12 ft 11½ in. (3.96 m)
Weight Empty: 6,543 lb (2,968 kg)
Weight Loaded: 9,921 lb (5,260 kg)
Max. Speed: 379 mph (329 kts)
Ceiling: 35,236 ft (10,470 m)
Endurance: Not known
Range: 1,914 miles (1,663 naut. m)
Engine: One 1,820-hp Nakajima NK9B Homare 11, 18-cylinder, air-cooled radial
                 One 1,990-hp Nakajima NK9H 21 Homare 21 18-cylinder,
                 air-cooled radial
                 One 1,980-hp Nakajima NK9K-L Homare 18-cylinder, air-cooled radial
Armament: One flexible 7.7 mm machine gun

Mitsubishi Ki-15 Army Type Reconnaissance (Babs)

Wing Span: 39 ft 4½ in. (12 m)
Length: 27 ft 10¼ in. (8.49 m) (Ki-15-I)
                 28 ft 6¾ in. (8.7 m) (Ki-15-II – C5M1&2)
Height: 10 ft 11½ in. (3.34 m) (Ki-15-I&II)
                 11 ft 4½ in. (3.46 m) (C5M1&2)
Weight Empty: 3,084 lb (1,399 kg) (Ki-15-I)
                 3,510 lb (1,592 kg) (Ki-15-II)
                 3,538 lb (1,605 kg) (C5M1)

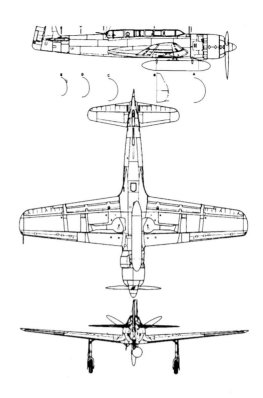

Nakajima C6N1 Saiun carrier
reconnaissance aircraft – Myrt.

	3,781 lb (1,715 kg) (C5M2)
Weight Loaded:	4,482 lb (2,033 kg) (Ki-15-I)
	4,826 lb (2,189 kg) (Ki-15-II)
	4,844 lb (2,197 kg) (C5M1)
	5,170 lb (2,345 kg) (C5M2)
Max. Speed:	298 mph (480 km/h) (Ki-15-I)
	317 mph (510 km/h) Ki-15-II)
	291 mph (253 kts) (C5M1)
	303 mph (263 kts) (C5M2)
Ceiling:	37,400 ft (11,400 m) (Ki-15-I&II)
	27,000 ft (8,230 m) (C5M1)
	31,430 ft (9,580 m) (C5M2)
Endurance:	Not known
Range:	1,491 miles (2,400 km) (Ki-15-I&II)
	725 miles (630 nautical miles) (C5M1)
	691 miles (600 nautical miles) (C5M2)
Engine:	One 550-hp Kakajima Ha-8 9-cylinder, air-cooled radial (Ki-15-I)
	One 900-hp Mitsubishi Ha-16-I, 14-cylinder, air-cooled radial (Ki-15-II)
	One 875-hp Mitsubishi Zuisei 14-cylinder, air-cooled radial (C5M1)
	One 950-hp Nakajima Sakae 12, 14-cylinder, air-cooled radial (C5M2)
Armament:	One flexible 7.7 mm machine gun (all models)

MITSUBISHI KI-30 ARMY TYPE LIGHT 97 BOMBER (ANN)

Wing Span:	47 ft 8½ in. (14.55 m)
Length:	33 ft 11¼ in. (10.34 m)
Height:	11 ft 11½ in. (3.64 m)
Weight Empty:	4,916 lb (2,230 kg)
Weight Loaded:	7,324 lb (3,322 kg)

Max. Speed: 263 mph (432 km/h)
Ceiling: 28,120 ft (8,570 m)
Endurance: Not known
Range: 1,056 miles (1,700 km)
Engine: One 825-hp Nakajima Ha-5, 14-cylinder, air-cooled radial
Armament: One wing mounted 7.7 mm machine gun
                One flexible rear-firing 7.7 mm machine gun
                Maximum bomb load of 882 lb (400 kg)

MITSUBISHI KI-21 ARMY TYPE 97 HEAVY BOMBER (SALLY)

Wing Span: 73 ft 9¾ in. (22.5 m) (Ia & IIb)
Length: 52 ft 5¾ in. (16 m) (Ia & IIb)
Height: 14 ft 3½ in. (4.35 m) (Ia & IIb)
Weight Empty: 10,342 lb (4,691 kg) (Ia)
                13,382 lb (6,070 kg) (IIb)
Weight Loaded: 16,517 lb (7,492 kg) (Ia)
                21,407 lb (9,710 kg) (IIa)
Max. Speed: 268 mph (432 km/h) (Ia)
                302 mph (486 km/h) (IIb)
Ceiling: 28,215 ft (8,600 m) (Ia)
                32,810 ft (10,000 m) (IIb)
Endurance: Not known
Range: 1,680 miles (2,700 km)
Engine: Two 850-hp Nakajima Ha-5, 14-cylinder, air-cooled radials (Ia)
                Two 1,450-hp Mitsubishi Ha-101, 14-cylinder, air-cooled radials (IIb)
Armament: One flexible 7.7 mm machine gun mounted either side of the fuselage
                One flexible 7.7 mm machine gun mounted in the nose, dorsal and ventral
                positions and one 7.7 mm machine gun mounted in the tail (Ia & IIb)
                Maximum bomb load of 2,205 lb (1,000 kg)

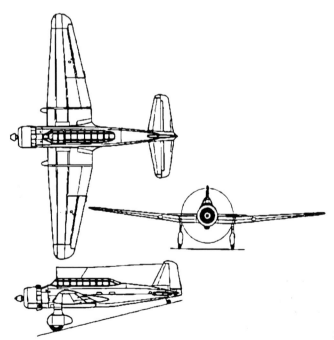

Mitsubishi Ki-30 Army light
bomber – Ann.

MITSUBISHI ARMY TYPE 99 KI-51 (SONIA)

Wing Span:	39 ft 8½ in. (12.1 m)
Length:	30 ft 2½ in. (9.21 m)
Height:	8 ft 11½ in. (2.73 m)
Weight Empty:	4,129 lb (1,873 kg)
Weight Loaded:	6,169 lb (2,798 kg)
Max. Speed:	263 mph (424 km/h)
Ceiling:	27,130 ft (8,270 m)
Endurance:	Not known
Range:	660 miles (1,060 km)
Engine:	One 900-hp Mitsubishi Ha-26-II, 14-cylinder, air-cooled radial
Armament:	Two wing-mounted 7.7mm machine guns
	One flexible rear-firing 7.7 mm machine gun
	Maximum bomb load of 441 lb (200 kg)

NAKAJIMA ARMY SPECIAL ATTACKER KI-115 TSURUGI (PEGGY)

Wing Span:	28 ft 2½ in. (8.6 m)
Length:	28 ft ½ in. (8.55 m)
Height:	10 ft 9½ in. (3.3 m)
Weight Empty:	3,616 lb (1,640 kg)
Weight Loaded:	5,688 lb (2,580 kg)
Max. Speed:	342 mph (550 km/h)
Ceiling:	21,325 ft (6,500 m)
Range:	745 miles (1,200 km)
Engine:	One 1,150-hp Nakajima Ha-35, 14-cylinder, air-cooled radial
Armament:	Maximum bomb load of 1,764 lb (800 kg)

KAWASAKI ARMY TYPE 1 FREIGHT TRANSPORT KI-56 (THALIA)

Wing Span:	65 ft 6 in. (19.96 m)
Length:	48 ft 10¾ in. (14.9 m)
Height:	11 ft 9¾ in. (3.6 m)
Weight Empty:	10,791 lb (4,895 kg)
Weight Loaded:	17,692 lb (8,025 kg)
Max. Speed:	248 mph (400 km/h)
Ceiling:	26,250 ft (8,000 m)
Endurance:	Not known
Range:	2,050 miles (1,782 naut. miles)
Engine:	Two 950-hp Nakajima Ha-25, 14-cylinder, air-cooled radials
Armament:	None

YOKOSHO NAVY TYPE 93 INTERMEDIATE TRAINER (K5Y1/2) (WILLOW)

Wing Span:	36 ft 1 in. (11 m) (K5Y1&2)
Length:	26 ft 4¾ in. (8.05 m) (K5Y1)
	28 ft 9¾ in. (8.78 m) (K5Y2)
Height:	10 ft 5¾ in. (3.2 m) (K5Y1)
	12 ft ¾ in. (3.68 m) (K5Y2)
Weight Empty:	2,205 lb (1,000 kg) (K5Y1)
	2,535 lb (1,150 kg) (K5Y2)
Weight Loaded:	3,307 lb (1,500 kg) (K5Y1)
	3,638 lb (1,650 kg) (K5Y2)

Max. Speed: 132 mph (115 knts) (K5Y1)
                   123 mph (107 knts) (K5Y2)
Ceiling: 18,700 ft (5,700 m) (K5Y1)
                   14,205 ft (4,330 m) (K5Y2)
Endurance: Not known
Range: 633 miles (550 naut. mls) (K5Y1)
                   436 miles (379 naut. mls)(K5Y2)
Engine: One 340-hp Hitachi Amikaze 11, 9-cylinder, air-cooled radial
                   One 515-hp Hitachi Amikaze 21, 9-cylinder, air-cooled radial
                   One 480-hp Hitachi Amikaze 21a, 9-cylinder, air-cooled radial
Armament: None

TACHIKAWA ARMY TYPE 95-1 INTERMEDIATE TRAINER KI-9 (SPRUCE)

Wing Span: 33 ft 10½ in. (10.32 m)
Length: 24 ft 8¼ in. (7.52 m)
Height: 9 ft 10¼ in. (3 m)
Weight Empty: 2,237 lb (1,015 kg)
Weight Loaded: 3,142 lb (1,425 kg)
Max. Speed: 149 mph (240 km/h)
Ceiling: 21,325 ft (6,500 m)
Range: 745 miles (1,200 km)
Endurance: 3½ hours
Engine: One 350-hp Hitachi Ha-13a, 9-cylinder, air-cooled radial
Armament: None

TACHIKAWA ARMY TYPE 93-3 PRIMARY TRAINER KI-17 (CEDAR)

Wing Span: 32 ft 2½ in. (9.82 m)
Length: 25 ft 7¼ in. (7.8 m)
Height: 9 ft 8¼ in. (2 m)
Weight Empty: 1,362 lb (618 kg)
Weight Loaded: 1,984 lb (900 kg)
Max. Speed: 106 mph (170 km/h)
Ceiling: 17,390 ft (5,300 m)
Range: 745 miles (1,200 km)
Endurance: 3½ hours.
Engine: One 350-hp Hitachi Ha-12, 7-cylinder, air-cooled radial

TACHIKAWA ARMY TYPE 1 TRAINER KI-54 (HICKORY)

Wing Span: 58 ft 8¾ in. (17.9 m)
Length: 39 ft 2¼ in. (11.9 m)
Height: 11 ft 9 in. (3.5 m)
Weight Empty: 6,512 lb (2,954 kg)
Weight Loaded: 3,897 lb (8,591 kg)
Max. Speed: 234 mph (376 km/h)
Ceiling: 22,555 ft (7,180 m)
Range: 597 miles (960 km)
Endurance: 3½ hours
Engine: Two 450-hp Hitachi Ha-13a, 9-cylinder, air-cooled radials
Armament: Four flexible 7.7 mm machine guns and practice bombs (Ki-54b)
                   1,058 lb (480 kg) depth charges (Ki-54d)

AICHI NAVY TYPE 99 CARRIER BOMBER D3A (VAL)

Wing Span:	47 ft 2 in. (14.37 m) (D3A1/2)
Length:	33 ft 5¼ in. (10.19 m) (D3A1/2)
Height:	12 ft 7½ in. (3.84 m) (D3A1/2)
Weight Empty:	5,309 lb (2,408 kg) (D3A1)
	5,666 lb (2,570 kg) (D3A2)
Weight Loaded:	8,047 lb (3,650 kg) (D3A1)
	8,378 lb (3,800 kg) D3A2)
Max. Speed:	240 mph (209 knt) (D3A1)
	267 mph (232 knt) (D3A2)
Ceiling:	30,050 ft (9,300 m) (D3A1)
	34,450 ft (10,500 m) (D3A2)
Range:	915 miles (795 naut/m) (D3A1)
	840 miles (730 naut/m) (D3A2)
Endurance:	Not known
Engine:	One 710-hp Nakajima Hikari 2, 9-cylinder, air-cooled radial (D3A 1st prototype)
	One 840-hp Mitsubishi Kinsei 3,14-cylinder, air-cooled radial (D3A 2nd prototype)
	One 1,000-hp Mitsubishi Kinsei 43,14-cylinder, air-cooled radial. (D3A1)
	One 1,300-hp Mitsubishi Kinsei 54, 14-cylinder, air-cooled radial. (D3A2)
Armament:	Four flexible 7.7 mm machine guns
	One 551 lb (250 kg) bomb under fuselage
	Two 132 lb (60 kg) bombs beneath the wings
Crew:	Two

AICHI NAVY TYPE O RECONNAISSANCE SEAPLANE E13A (JAKE)

Wing Span:	58 ft 8¾ in. (17.9 m)
Length:	39 ft 2¼ in. (11.9 m)
Height:	11 ft 9 in. (3.5 m)
Weight Empty:	6,512 lb (2,954 kg)
Weight Loaded:	3,897 lb (8,591 kg)
Max. Speed:	234 mph (376 km/h)
Ceiling:	22,555 ft (7,180 m)
Range:	597 miles (960 km)
Endurance:	3½ hours
Engine:	Two 450-hp Hitachi Ha-13a, 9-cylinder, air-cooled radials
Armament:	Four flexible 7.7mm machine guns and practice bombs (Ki-54b)
	1,058 lb (480 kg) depth charges (Ki-54d)

MITSUBISHI EXPERIMENTAL INTERCEPT FIGHTER KI-109

Wing Span:	73 ft 9¾ in. (22.5 m)
Length:	58 ft 10¼ in. (17.95 m)
Height:	19 ft 1 in. (5.8 m)
Weight Empty:	16,367 lb (7,424 kg)
Weight Loaded:	23,810 lb (10,800 kg)
Max. Speed:	342 mph (550 km/h)
Ceiling:	31,070 ft (9,470 m)
Range:	1,367 miles (2,200 km)
Endurance:	Not known
Engines:	Two 1,900-hp Hitachi Ha-104, 18-cylinder, air-cooled radials
Armament:	One forward-firing 75 mm Type 88 cannon

One flexible 12.7 mm machine gun mounted in rear turret

AICHI NAVY RECONNAISSANCE SEAPLANE E16A1 (PAUL)

Wing Span: 42 ft ½ in. (12.8 m)
Length: 35 ft 6½ in. (10.83 m)
Height: 15 ft 8¾ in. (4.79 m)
Weight Empty: 6,493 lb (2,945 kg)
Weight Loaded: 8,598 lb (3,900 kg)
Max. Speed: 273 mph (237 knts)
Ceiling: 32,810 ft (10,000 m)
Range: 1,504 miles (1,307 naut. miles)
Endurance: Not known
Engines: One 1,300-hp Mitsubishi MK8A Kinsei 51, 14-cylinder, air-cooled radial
Armament: Two wing mounted 20 mm cannons
                 One flexible 7.7 mm rear-firing machine gun

AICHI NAVY CARRIER ATTACK BOMBER B7A1 (GRACE)

Wing Span: 47 ft 2½ in. (14.4 m)
Length: 37 ft 8½ in. (11.49 m)
Height: 13 ft 4½ in. (4.07 m)
Weight Empty: 8,400 lb (3,810 kg)
Weight Loaded: 12,401 lb (5,625 kg)
Max. Speed: 352 mph (306 knts)
Ceiling: 36,910 ft (11,250 m)
Range: 1,640 miles (1,888 naut. miles)
Endurance: Not known
Engines: One 1,800-hp Nakajima NK9B Homare 11, 18-cylinder, air-cooled radial
                 One 1,825-hp Nakajima NK9C Homare 12, 18-cylinder, air-cooled radial
Armament: Two wing mounted 20 mm cannons
                 One flexible 7.7 mm rear-firing machine gun
Crew: Two

MITSUBISHI ARMY TYPE 4 HEAVY BOMBER KI-67-I HIRYU (FLYING DRAGON)

Wing Span: 73 ft 9¾ in. (22.5 m)
Length: 61 ft 4¼ in. (18.7 m)
Height: 25ft 3¼ in. (7.7 m)
Weight Empty: 19,068 lb (8,649 kg)
Weight Loaded: 30,347 lb (13,765 kg)
Max. Speed: 334 mph (537 km/h)
Ceiling: 31,070 ft (9,470 m)
Range: 2,360 miles (3,800 km)
Endurance: Not known
Engines: Two 2,400-hp Mitsubishi Ha-214, 18-cylinder, air-cooled radials
                 Two 1,825-hp Mitsubishi Ha-104, 18-cylinder, air-cooled radials
Armament: One flexible 7.92mm machine guns mounted in the nose, port and
                 starboard blisters
                 Twin flexible 12.7 mm machine guns mounted in tail turret
                 One 20 mm cannon mounted in the dorsal turret
Crew: Six to eight

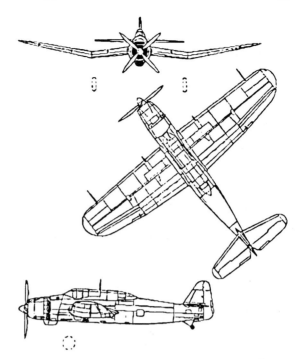

Aichi B7A1 Navy carrier attack
bomber – Grace

Kawanishi Navy Type 94 Reconnaissance Seaplane (E7K1/2) (Alf)

Wing Span:	45 ft 11½ in. (14 m) (E7K1&2)
Length:	34 ft 1¼ in. (10.4 m) (E7K1&2)
Height:	15 ft 9¼ in. (4.8 m) (E7K1&2)
Weight Empty:	4,343 lb (1,970 kg) (E7K1)
	4,630 lb (2,100 kg) (E7K2)
Weight Loaded:	6,614 lb (3,000 kg) (E7K1)
	7,725 lb (3,300 kg) (E7K2)
Max. Speed:	148 mph (129 knts) E7K1)
	171 mph (149 knts) (E7K2)
Ceiling:	23,165 ft (7,060 m) (E7K1&2)
Range:	Not known
Endurance:	12 hours (E7K1&2)
Engines:	One 500-hp Hiro Type 91, 12-cylinder, liquid-cooled. (E7K1)
	One 600-hp Hiro Type 91, 12-cylinder, liquid-cooled (E7K1)
	One 870-hp Mitsubishi Zuisei 11, 14-cylinder, air-cooled, radial (E7K2)
Armament:	One forward-firing 7.7 mm machine gun
	One flexible rearward-firing 7.7 mm machine gun
	One downward-firing 7.7 mm machine gun

Kyushu Navy Operations Trainer K11W Shiragiku

Wing Span:	49 ft 1¾ in. (14.98 m)
Length:	33 ft 7 in. (10.24 m)
Height:	12 ft 10¾ in. (3.93 m)
Weight Empty:	3,697 lb (1,677 kg)
Weight Loaded:	5,820 lb (2,640 kg)
Max. Speed:	143 mph (124 knts)
Ceiling:	18,440 ft (5,620 m)

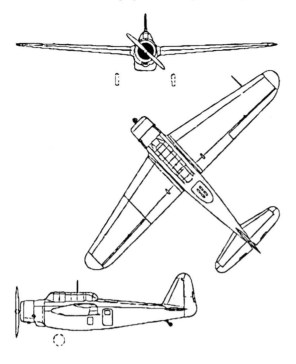

Kyushu K11W Shiragiku Navy
operations trainer.

Range: 1,093 miles (950 nm)
Endurance: Not known
Engines: One 515-hp Hitachi GK2B Amakaze 21, 9-cylinder, air-cooled radial
Armament: One flexible 7.7 mm rear-firing machine gun
Crew: Five

KAWANISHI NAVY TYPE 97 RECONNAISSANCE FLYING BOAT (H6K)

Wing Span: 131 ft 2¾ in. (40 m) (H6K2/4/5)
Length: 84 ft 1 in. (25.62 m) (H6K2/4/5))
Height: 20 ft 6¼ in. (6.27 m) (H6K2/4/5)
Weight Empty: 22,796 lb (10,340 kg) (H6K2)
                 26,511 lb (12,025 kg) (H62-L)
                 25,810 lb (11,707 kg) (H6K4)
                 27,117 lb (12,380 kg) (H6K5)
Weight Loaded: 35,274 lb (16,000 kg) (H6K2)
                 37,699 lb (17,100 kg) (H6K2-L)
                 37,479 lb (21,500 kg) H6K4
                 38,581 lb (23,000 kg) H6K5
Max. Speed: 206 mph (179 knts) (H6K2/2-L/4)
                 239 mph (208 knts) (H6K5)
Ceiling: 24,935 ft (7,600 m) (H6K2)
                 31,365 ft (9,610 m) (H6K4/5)
Range: 2,567 miles (2,230 nm) (H6K2)
                 2,690 miles (2,337 nm) (H6K2-L)
                 3,779 miles (3,283 nm) (H6K4)
                 4,208 miles (3,656 nm) (H6K5)
Endurance: Not known
Engines: Four 840-hp Nakajima Hikari 2, 9-cylinder, air-cooled radials. (H6K1)
                 Four 1,000-hp Mitsubishi Kinsei 46, 14-cylinder, air-cooled radials

(H6K1/3/4/5)
Four 1,300-hp Mitsubishi Kinsei 51, 14-cylinder, air-cooled, radial (H6K4)

Armament:	One flexible 7.7 mm machine gun in bow position/forward turret
	One flexible 7.7 mm machine gun in dorsal position, one 7.7 mm machine gun in blister either side of fuselage
	One hand-held 7.7 mm machine gun in tail turret
	2,205 lb (1,000 kg) bombs or two 1,764 lb (800 kg) torpedoes

KAWANISHI NAVY TYPE 2 RECONNAISSANCE SEAPLANE E15K1 SHIUN (NORM)

Wing Span:	45 ft 11¼ in. (14 m)
Length:	38 ft ¼ in. (11.58 m)
Height:	16 ft 2¾ in. (4.95 m)
Weight Empty:	6,978 lb (3,165 kg)
Weight Loaded:	9,039 lb (4,100 kg)
Max. Speed:	291 mph (253 knts)
Ceiling:	32,250 ft (9,830 m)
Range:	2,095 miles (1,820 nm)
Endurance:	Not known
Engines:	One 1,500-hp Mitsubishi MK4D Kasei 14, 14-cylinder, air-cooled radial
	One 1,850-hp Mitsubishi MK4S Kasei 24, 14-cylinder, air-cooled radial
Armament:	One flexible 7.7 mm rear-firing machine gun
Crew:	Two

WATANABE (KYUSHU) NAVY PATROL Q1W (LORNA)

Wing Span:	52 ft ½ in. (16 m)
Length:	39 ft 7¾ in. (12.08 m)
Height:	13 ft 6¼ in. (4.11 m)
Weight Empty:	6,839 lb (3,102 kg)
Weight Loaded:	10,582 lb (4,800 kg)
Max. Speed:	200 mph (174 knts)
Ceiling:	14,730 ft (4,490 m)
Range:	834 miles (725 naut. miles)
Endurance:	Not known
Engines:	Two 610-hp Hitachi GK2C Amakaze 31, 9-cylinder, air-cooled radials

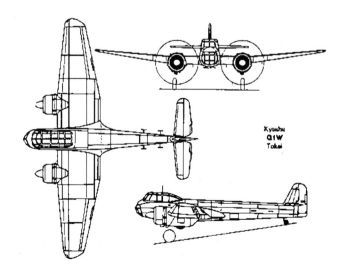

Watanabe Q1W Navy patrol
aircraft – Lorna.

Armament: One flexible 7.7 mm rear-firing machine gun
                 Two forward firing 20 mm cannons
Crew: Three – Four

WATANABE NAVY TYPE 96 SMALL RECONNAISSANCE SEAPLANE (E9W1) (SLIM)

Wing Span: 32 ft 9 in. (9.98 m)
Length: 25 ft 0¾ in. (7.64 m)
Height: 10 ft 9½ in. (3.3 m)
Weight Empty: 1,876 lb (847 kg)
Weight Loaded: 2,667 lb (1,210 kg)
Max. Speed: 145 mph (126 knts)
Ceiling: 22,112 ft (6,740 m)
Range: 454 miles (395 naut. miles)
Endurance: 4.9 hours
Engines: One 340-hp Gasuden Tempu 11, 9-cylinder, air-cooled radials
Armament: One dorsal flexible 7.7 mm rear-firing machine gun
Crew: Two

YOKOSUKA MXY7 NAVY SUICIDE ATTACK OHKA (CHERRY BLOSSOM)

Wing Span: 16 ft 9½ in. (5.12 m) (Model 11)
                 13 ft 6½ in. (4.12 m) (Model 22)
Length: 19 ft 10¾ in. (6.06 m) (Model 11)
                 22 ft 6¾ in. (6.88 m) (Model 22)
Height: 3 ft 9¾ in. (1.16 m) (Model 11 & 22)
Weight Empty: 970 lb (440 kg) (Model 11)
                 1,202 lb (545 kg) (Model 22)
Weight Loaded: 4,718 lb (2,140 kg) (Model 11)
                 3,197 lb (1,450 kg) (Model 22)
Max. Speed: 403 mph (350 knts) (Model 11)
                 276 mph (240 knts) (Model 22)
Range: 23 miles (20 naut. miles) (Model 11)
                 81 miles (70 naut. miles) (Model 22)
Engines: Three Type 4, Mark 1 Model 20 solid-propellant rockets
                 One 550 lb (200 kg) Ne-20 axial-flow turbo-jet
Armament: 2,646 lb (1,200 kg) warhead
Crew: One

YOKOSUKA (KUGISHO) NAVY TYPE O SMALL RECONNAISSANCE SEAPLANE E14Y1 (GLEN)

Wing Span: 36 ft 1½ in. (11 m)
Length: 28 ft ¼ in. (8.54 m)
Height: 12 ft 5¼ in. (3.8 m)
Weight Empty: 2,469 lb (1,119 kg)
Weight Loaded: 3,197 lb (1,450 kg)
Max. Speed: 153 mph (133 knts)
Ceiling: 17,780 ft (5,420 m)
Range: 548 miles (476 naut. miles)
Endurance: Not known
Engines: One 340-hp Hitachi Tempu 12, 9-cylinder, air-cooled radial
Armament: One flexible 7.7 mm rear-firing machine gun

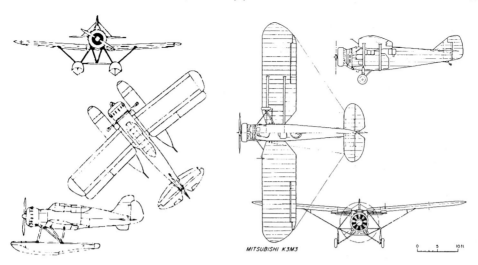

MITSUBISHI K3M3

Left: Yokusuka E14Y reconnaissance seaplane – Glen.
Right: Mitsubishi K3M1-3 Navy crew trainer – Pine.

Crew: Two

MITSUBISHI NAVY TYPE 90 CREW TRAINER K3M1-3 (PINE)

Wing Span:	51 ft 9¼ in. (15.78 m)
Length:	31 ft 3½ in. (9.54 m)
Height:	12 ft 6½ in. (3.82 m)
Weight Empty:	2,998 lb (1,360 kg)
Weight Loaded:	4,850 lb (2,200 kg)
Max. Speed:	146 mph (127 knts)
Ceiling:	20,965 ft (6,390 m)
Range:	497 miles (432 naut. miles)
Endurance:	Not known
Engines:	One 340-hp Mitsubishi-built Hispano Suiza, 8-cylinder, liquid-cooled
	One 340-hp Hitachi Amakaze 11, 9-cylinder, air-cooled radial
	One 580-hp Nakajima Kotobuki 2 KA12, 9-cylinder, air-cooled radial
Armament:	One flexible 7.7mm rear-firing machine gun
	Four 66 lb (30 kg) bombs
Crew:	Two

MITSUBISHI NAVY TYPE 96 ATTACK BOMBER G3M1 (NELL)

Wing Span:	82 ft ¼ in. (25 m) (G3M1&2)
Length:	53 ft 11¾ in. (16.45 m) (G3M1&2)
Height:	12 ft 1 in. (3.68 m) (G3M1&2)
Weight Empty:	10,516 lb (4,770 kg) (G3M1)
	10,936 lb (4,965 kg) (G3M2)
Weight Loaded:	16,848 lb (7,642 kg) (G3M1)
	17,637 lb (8,000 kg) (G3M2)
Max. Speed:	216 mph (188 knts) (G3M1)
	232 mph (201 knts) (G3M2)

Ceiling:	25,540 ft (7,480 m) (G3M1)
	29,950 ft (9,130 m) (G3M2)
Range:	2,722 miles (2365 naut. mls) (G3M1&2)
Endurance:	Not known
Engines:	Two 1,075-hp Mitsubishi Kinsei 41, 14-cylinder, air-cooled radial.
	Two 1,300-hp Mitsubishi Kinsei 45. 14-cylinder, air-cooled radials
Armament:	One flexible 7.7 mm machine gun mounted in each blister either side of the fuselage.
	One 7.7 mm machine gun mounted in retractable dorsal turret and one 20 mm cannon mounted in fixed dorsal turret
Crew:	Five – seven

MITSUBISHI NAVY TYPE O OBSERVATION SEAPLANE F1M2 (PETE)

Wing Span:	36 ft 1¼ in. (11 m)
Length:	31 ft 2 in. (9.5 m)
Height:	13 ft 1½ in. (4 m)
Weight Empty:	4,251 lb (1,928 kg)
Weight Loaded:	5,622 lb (2,550 kg)
Max. Speed:	230 mph (200 knts)
Ceiling:	30,970 ft (9,440 m)
Range:	460 miles (400 naut. miles)
Engines:	One 875-hp Mitsubishi Zuisei 13, 14-cylinder, air-cooled radial
Armament:	Two fixed forward-firing 7.7 mm machine guns mounted in the upper section of the fuselage
	One rear-firing 7.7 mm machine gun
Crew:	Two – pilot and observer/radio operator

MITSUBISHI NAVY TYPE 96 CARRIER FIGHTER A5M (CLAUDE)

Wing Span:	36 ft 1¼ in. (11 m)
Length:	25 ft 1¾ in. (7.67 m)
Height:	10 ft 8½ in. (3.26 m)
Weight Empty:	2,370 lb (1,075 kg)
Weight Loaded:	3,307 lb (1,500 kg)
Max. Speed:	280 mph (243 knts)
Ceiling:	32,150 ft (9,800 m)
Range:	746 miles (648 naut. miles)
Endurance:	Not known
Engines:	One 640-hp Nakajima Kotobuki 3, 9-cylinder, air-cooled radial
Armament:	Two fixed forward-firing 7.7 mm machine guns mounted in the upper section of the fuselage
Crew:	One

KAWASAKI ARMY TYPE 4 ASSAULT KI-102 (RANDY)

Wing Span:	51 ft 1 in. (15.57 m) (Ki-102b)
	56 ft 7½ in. (17.25 m) (Ki-102c)
Length:	37 ft 6½ in. (11.45 m) (Ki-102b)
	42 ft 9¾ in. (13.05 m) (Ki-102c)
Height:	12 ft 1¾ in. (3.70 m) (Ki-102b&c)
Weight Empty:	10,913 lb (4,950 kg) (Ki-102b)
	11,464 lb (5,200 kg) (Ki-102c)
Weight Loaded:	16,094 lb (7,300 kg) (Ki-102b)

16,755 lb (7,600 kg) (Ki-102c)
Max. Speed: 360 mph (580 km/h) (Ki-102b)
                 373 mph (600 km/h) (Ki-102c)
Ceiling: 32,810 ft (10,000 m) (Ki-102b)
                 44,290 ft (13,500 m) (Ki-102c)
Range: 1,243 miles (2,000 km) (Ki-102b)
                 1,367 miles (2,200 km) (Ki-102c)
Engines: Two 1,500-hp Mitsubishi Ha-112, 14-cylinder, air-cooled radials
Armament: One 57 mm cannon in the nose
                 Two 20 mm cannons mounted in the belly of the fuselage
                 One rearward firing flexible 12.7 mm machine gun (Ki-102b)
                 Two 30 mm cannons mounted in the belly of the fuselage and two
                 20 mm cannons mounted obliquely in the fuselage behind the cockpit
Crew: Two

KAWASAKI ARMY TYPE 99 LIGHT BOMBER KI-48 (LILY)

Wing Span: 57 ft 3¾ in. (17.47 m)
Length: 41 ft 4¼ in. (12.6 m)
Height: 12 ft 5¾ in. (3.8 m)
Weight Empty: 8,929 lb (4,050 kg)
Weight Loaded: 13,007 lb (5,900 kg)
Max. Speed: 298 mph (480 km/h)
Ceiling: 31,170 ft (9,500 m)
Range: 1,230 miles (1,980 km)
Engines: Two 1,150-hp Nakajima Ha-115, 14-cylinder, air-cooled radials
Armament: Two 7.7 mm machine guns in the nose
                 One 7.7 mm machine gun mounted in a ventral position
Crew: Four

MITSUBISHI NAVY TYPE O CARRIER FIGHTER A6M REISEN (ZEKE)

Wing Span: 39 ft 4½ in. (12 m) (A6M2)
                 36 ft 1¼ in. (11 m) (A6M3/5)
Length: 29 ft 8¾ in. (9.06 m) (A6M2/3)
                 29 ft 11¼ in. (9.12 m) (A6M5)
Height: 10 ft ¼ in. (3.05 m) (A6M2)
                 11 ft 6¼ in. (3.5 m) (A6M3/5)
Weight Empty: 3,704 lb (1,680 kg) (A6M2)
                 3,984 lb (1,807 kg) (A6M3)
                 4,136 lb (1,876 kg) (A6M5)
Weight Loaded: 5,313 lb (2,410 kg) (A6M2)
                 5,609 lb (2,544 kg) (A6M3)
                 6,025 lb (2,733 kg) (A6M5)
Max. Speed: 331 mph (288 knts) (A6M2)
                 338 mph (294 knts) (A6M3)
                 351 mph (305 knts) (A6M5)
Ceiling: 32,810 ft (10,000 m) (A6M2)
                 36,250 ft (11,050 m) (A6M3)
                 38,520 ft (11,740 m) (A6M5)
Range: 1,930 miles (1,675 naut. mls) (A6M2)
                 1,477 miles (1,284 naut. mls) (A6M3)
                 1,194 miles (1,037 naut. mls) (A6M5)
Engines: One 780-hp Nakajima Sakae 12, 14-cylinder, air-cooled radial(A6M2)

One 1,100-hp Nakajima Sakae 21, 14-cylinder, air-cooled radial (A6M3/5)

Armament: Two fixed forward-firing 7.7 mm machine guns mounted in the upper
               section of the fuselage
               Two wing mounted 20 mm cannons (A6M2/3/5)
Crew: One

AICHI M6A SEIRAN

Wing Span: 40 ft 2¾ in. (12.6 m)
Length: 38 ft 2¼ in. (11.6 m)
Height: 15 ft ½ in. (4.58 m)
Weight Empty: 7,277 lb (3,301 kg)
Weight Loaded: 8,907 lb (4,040 kg)
Max. Speed: 295 mph (256 knts)
Ceiling: 32,480 ft (9,900 m)
Range: 739 miles (642 naut. mls)
Engines: One 1,400-hp Aichi Atsuta 31, 12-cylinder, vee liquid-cooled
Armament: One flexible rear-firing 13 mm machine gun
Crew: Crew in tandem

KAWASAKI KI-61 HIEN (TONY)

Wing Span: 39 ft 4½ in. (12 m)
Length: 28 ft 8½ in. (8.7 m)
Height: 12 ft 1½ in. (3.7 m)
Weight Empty: 4,872 lb (2,210 kg)
Weight Loaded: 6,504 lb (2,950 kg)
Max. Speed: 368 mph (592 km/h)
Ceiling: 37,730 ft (11,600 m)
Range: 373 miles (600 km)
Engines: One 1,500-hp Kawasaki Ha-140, 12-cylinder, Vee liquid-cooled
Armament: Two fuselage mounted 20 mm cannons
               Two wing-mounted 12.7 mm machine guns
Crew: One

KAWASAKI KI-100 ARMY TYPE 5 FIGHTER

Wing Span: 39 ft 4½ in. (12 m) (I&II)
Length: 28 ft 11¼ in. (8.8 m) (I&II)
Height: 12 ft 3½ in. (3.75 m) (I&II)
Weight Empty: 5,567 lb (2,525 kg) (I)
               5,952 lb (2,700 kg) (II)
Weight Loaded: 7,705 lb. (3,495 kg) (I)
               8,091 lb (3,670 kg) (II)
Max. Speed: 360 mph (580 km/h) (I)
               354 mph (570 km/h) (II)
Ceiling: 36,090 ft (11,000 m) (I&II)
Range: 870 miles (1,400 km.) (I&II)
Engines: One 1,500-hp Mitsubishi Ha-33, 14-cylinder, air-cooled
Armament: Two fuselage mounted 20 mm cannons
               Two wing-mounted 12.7 mm machine guns
Crew: One

Mitsubishi Ki-46 Army Type Reconnaissance (Dinah)

Wing Span:	48 ft 2¾ in. (14.7 m) (I, II, III & IV)
Length:	36 ft 1¼ in. (11 m) (I, II, III & IV)
Height:	12 ft 8¾ in. (3.88 m) (I, II, III & IV)
Weight Empty:	7,449 lb (3,379 kg) (I)
	7,194 lb (3,263 kg) (II)
	8,446 lb (3,831 kg) (III)
	8,840 lb (4,010 kg) (IV)
Weight Loaded:	10,631 lb. (4,822 kg) (I)
	11,133 lb (5,050 kg) (II)
	12,619 lb (5,722 kg) (III)
	13,007 lb (5,900 kg) (IV)
Max. Speed:	336 mph. (540 km/h) (I)
	375 mph (604 km/h) (II)
	391 mph (630 km/h) (III & IV)
Ceiling:	35,530 ft (10,830 m) (I)
	35,170 ft (10,720 m) (II)
	34,450 ft (10,500 m) (III)
	36,090 ft (11,000 m) (IV)
Range:	1,305 miles (2,100 km) (I)
	1,537 miles (2,474 km) (II)
	2,485 miles (4,000 km) (III & IV)
Engines:	Two 900-hp Mitsubishi Ha-26, 14-cylinder, air-cooled radials (I)
	Two 1,050-hp Mitsubishi Ha-102, 14-cylinder, air-cooled radials (II)
	Two 1,500-hp Mitsubishi Ha-112, 14-cylinder, air-cooled radials (III & IV)
Armament:	One obliquely 37 mm cannon mounted in the fuselage
	Two nose mounted 20 mm cannons
Crew:	One

Mitsubishi Ki-57 Army Type 100 Transport (Topsy)

Wing Span:	74 ft 1¾ in. (22.6 m) (I & II)
Length:	52 ft 9¾ in. (16.1 m) (I & II)
Height:	15 ft 7¾ in. (4.77 m) (I)
	15 ft 11½ in. (4.86 m) (II)
Weight Empty:	12,174 lb (5,522 kg) (I)
	12,313 lb (5,585 kg) (II)
Weight Loaded:	17,328 lb (7,860 kg) (I)
	18,018 lb (8,173 kg) (II)
Max. Speed:	267 mph (430 km/h) (I)
	292 mph (470 km/h) (II)
Ceiling:	22,965 ft (7,000 m) (I)
	26,250 ft (8,000 m) (II)
Range:	932 miles (1,500 km) (I & II)
Engines:	Two 850-hp Nakajima Ha-5, 14-cylinder, air-cooled radials (I)
	Two 1,050-hp Mitsubishi Ha-102, 14-cylinder, air-cooled radials (II)
Armament:	None
Crew:	Four with accommodation for eleven troops

Nakajima Ki-44 Army Type 2 Fighter Shoki (Tojo)

Wing Span:	31 ft ¼ in. (9.45 m) (Ia & IIb)
Length:	28 ft 8½ in. (8.75 m) (Ia)

	28 ft 9¾ in. (8.78 m) (IIb)
Height:	10 ft 8 in. (3.25 m) (Ia & IIb)
Weight Empty:	4,286 lb (1,944 kg) (Ia)
	4,643 lb (2,106 kg) (IIb)
Weight Loaded:	5,622 lb (2,550 kg) (Ia)
	6,094 lb (2,764 kg) (IIb)
Max. Speed:	360 mph (580 km/h) (Ia)
	376 mph (605 km/h) (IIb)
Ceiling:	35,500 ft (10,820 m) (Ia)
	36,745 ft (11,200 m) (IIb)
Range:	1,070 miles (1,722 km) (Ia & IIb)
Engines:	One 1,250-hp Nakajima Ha-41, 14-cylinder, air-cooled radial (Ia)
	One 2,000-hp Mitsubishi Ha-145, 18-cylinder, air-cooled radial (IIb)
Armament:	Two 20 mm cannons mounted on the fuselage
	Two wing-mounted 12.7 mm machine guns (Ia & IIb)

NAKAJIMA KI-84 ARMY TYPE 2 FIGHTER HAYATE (FRANK)

Wing Span:	36 ft 10¼ in. (11.23 m)
Length:	32 ft 6½ in. (9.92 m)
Height:	11 ft 1¼ in. (3.38 m)
Weight Empty:	5,864 lb (2,660 kg)
Weight Loaded:	7,955 lb (3,613 kg)
Max. Speed:	392 mph (631 km/h)
Ceiling:	34,450 ft (10,500 m)
Range:	1,053 miles (1,695 km)
Engines:	One 1,500-hp Mitsubishi Ha-33, 14-cylinder, air-cooled radial
Armament:	Two 12.7 mm machine guns mounted on the fuselage
	Two wing-mounted 20 mm cannons

KAWANISHI H8K (EMILY)

Wing Span:	124 ft 8¼ in. (38 m) (H8K1, 2, 3 & L)
Length:	92 ft 3¾ in. (28.13 m) (H8K1, 2, 3 & L)
Height:	30 ft 0¼ in. (9.15 m) (H8K1, 2, 3 & L)
Weight Empty:	34,176 lb (15,502 kg) (H8K1)
	40,521 lb (18,380 kg) (H8K2)
	40,940 lb (18,570 kg) (H8K3)
	37,258 lb (16,900 kg) (H8KL)
Weight Loaded:	54,013 lb (24,500 kg) (H8K1, 2 & 3)
	58,286 lb (26,683 kg) (H8KL)
Max. Speed:	269 mph. (234 knts) (H8K1)
	290 mph (252 knts) (H8K2 & 3)
	261 mph (227 knts) (H8KL)
Ceiling:	25,035 ft (7,630 m) (H8K1)
	29,035 ft (8,850 m) (H8K2, 3 & L)
Range:	4,475 miles (3,888 naut. mls) (H8K1)
	4,445 miles (3,862 naut. mls) (H8K2 & 3)
	2,759 miles (2,397 naut. mls) (H8KL)
Engines:	Four 1,530-hp Mitsubishi MK4B Kasei 12, 14-cylinder, air-cooled radials (H8K1)
	Four 1,850-hp Mitsubishi MK4Q Kasei 22, 14-cylinder, air-cooled radials
Armament:	One 20 mm cannon mounted in the bow, dorsal and tail turrets (H8K1)
	One 20mm cannon mounted in the dorsal and tail turrets

Two 7.7 mm machine guns mounted in beam blisters either side of the fuselage and cockpit hatches (H8K2, 3 & L)

Crew: Ten

KAWANISHI N1K1/2-J SHIDEN

Wing Span:	39 ft 4½ in. (12 m) (N1K1 & 2)
Length:	29 ft 1¾ in. (8.88 m) (N1K1)
	30 ft 7¾ in. (9.34 m) (N1K2)
Height:	13 ft 3¾ in. (4.06 m) (N1K1)
	12 ft 11¾ in. (3.96 m) (N1K2)
Weight Empty:	6,387 lb (2,897 kg) (N1K1)
	5,858 lb (2,657 kg) (N1K2)
Weight Loaded:	8,598 lb (3,900 kg) (N1K1)
	8,818 lb (4,000 kg) (N1K2)
Max. Speed:	363 mph (315 knts) (N1K1)
	369 mph (321 knts) (N1K2)
Ceiling:	41,010 ft (12,500 m) (N1K1)
	35,300 ft (10,760 m) (N1K2)
Range:	890 miles (773 naut. mls) (N1K1)
	1,066 miles (926 naut. mls) (N1K2)
Engines:	One 2,000-hp Nakajima NK9H, 18-cylinder, air-cooled radial (N1K1 & 2)
Armament:	Two 7.7 mm machine guns mounted on the fuselage
	Four wing-mounted 20 mm cannons
Crew:	One

MITSUBISHI J2M RAIDEN (JACK)

Wing Span:	35 ft 5¼ in. (10.8 m) (J2M1, 2, 3, 4, & 5)
Length:	32 ft 5¾ in. (9.9 m) (J2M1)
	31 ft 9¾ in. (9.69 m) (J2M2)
	32 ft 7½ in. (9.94 m) (J2M3)
	33 ft 3½ in. (10.14 m) (J2M4)
	32 ft 7½ in. (9.94 m) (J2M5)
Height:	12 ft 6¾ in. (3.82 m) (J2M1)
	12 ft 8¼ in. (3.87 m) (J2M2)
	12 ft 11¼ in. (3.96 m) (J2M3, 4 & 5)
Weight Empty:	4,830 lb (2,191 kg) (J2M1)
	5,176 lb (2,348 kg) (J2M2)
	5,423 lb (2,460 kg) (J2M3)
	6,202 lb (2,823 kg) (J2M4)
	5,534 lb (2,510 kg) (J2M5)
Weight Loaded:	6,307 lb (2,867 kg) (J2M1)
	7,077 lb (3,210 kg) (J2M2)
	7,573 lb (3,435 kg) (J2M3)
	8,702 lb (3,947 kg) (J2M4)
	7,767 lb (3,482 kg) (J2M5)
Max. Speed:	359 mph (312 knts) (J2M1)
	371 mph (322 knts) (J2M2)
	365 mph (317 knts) (J2M3)
	362 mph (315 knts) (J2M4)
	382 mph (332 knts) (J2M5)
Ceiling:	36,090 ft (11,000 m) (J2M1&2)
	38,385 ft (11,700 m) (J2M3)

	37,895 ft (11,550 m) (J2M4)
	36,910 ft (11,250 m) (J2M5)
Range:	1,180 miles (1,025 naut. mls) (J2M1, 2 & 3)
	575 miles (500 naut. mls) (J2M4)
	783 miles (680 naut. mls) (J2M5)
Engines:	One 1,800-hp Mitsubishi MK4R, 14-cylinder, air-cooled radial (J2M1, 2, 3, 4 & 5)
Armament:	Two 7.7 mm machine guns mounted on the fuselage
	Two/four wing-mounted 20 mm cannons

MITSUBISHI J8M SHSUI

Wing Span:	31 ft 2 in. (9.5 m)
Length:	19 ft 10¼ in. (6.05 m)
Height:	8 ft 10¼ in. (2.7 m)
Weight Empty:	3,318 lb (1,505 kg)
Weight Loaded:	8,565 lb (3,885 kg)
Max. Speed:	559 mph (900 km/h)
Ceiling:	39,370 ft (12,000 m)
Range:	Not known
Engines:	One 3,370 lb (1,500 kg) thrust Toko R2 bi-fuel liquid rocket motor
Armament:	Two wing-mounted 30 mm cannons
Crew:	One

NAKAJIMA J1N GEKKO (IRVING)

Wing Span:	55 ft 8½ in. (16.9 m) (J1N1/C/S)
Length:	39 ft 11¾ in. (12.18 m) (J1N1/C)
	41 ft 10¼ in. (12.7 m) (J1N1-S)
Height:	14 ft 11½ in. (4.56 m) (J1N1/C/S)
Weight Empty:	11,067 lb (5,020 kg) (J1N1)
	10,697 lb (4,852 kg) (J1N1-C)
	10,670 lb (4,840 kg) (J1N1-S)
Weight Loaded:	15,984 lb (8,030 kg) (J1N1)
	15,190 lb (7,527 kg) (J1N1-C)
	15,454 lb (7,010 kg) (J1N1-S)
Max. Speed:	315 mph (274 knts) (J1N1)
	329 mph (286 knts) (J1N1-C)
	315 mph (274 knts) (J1N1-S)
Ceiling:	33,795 ft (10,300 m) (J1N1/C)
	30,610 ft (9,320 m) (J1N1-S)
Range:	1,677 miles (1,457 naut. mls) (J1N1/C)
	1,581 miles (1,374 naut. mls) (J1N1-S)
Engines:	Two 1,130-hp Nakajima NK1F, 14-cylinder, air-cooled radial. (J1N1/C/S)
Armament:	One forward-firing 20 mm cannon
	Two forward-firing 7.7 mm machine guns
	Four 7.7 mm machine guns mounted in two remote-controlled dorsal barbettes
Crew:	Three

YOKOSUKA D4Y SUISEI (JUDY)

Wing Span:	37 ft 8¼ in. (11.5 m) (D4Y1/2/3/4)
Length:	33 ft 6¼ in. (10.2 m) (D4Y1/2/3/4)
Height:	12 ft 11½ in. (3.67 m) (D4Y1/2/3/4)

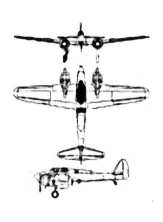
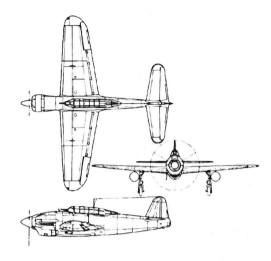

Above, left: Nakajima J1N Gekko – Irving.
Above, right: Yokosuka D4Y Suisei – Judy.

Weight Empty:	5,379 lb (2,440 kg) (D4Y1)
	5,809 lb (2,635 kg) (D4Y2)
	5,514 lb (2,501 kg) (D4Y3)
	5,809 lb (2,635 kg) (D4Y4)
Weight Loaded:	8,047 lb (3,650 kg) (D4Y1)
	8,455 lb (3,835 kg) (D4Y2)
	8,276 lb (3,754 kg) (D4Y3)
	10,013 lb (4,542 kg) (D4Y4)
Max. Speed:	343 mph (298 knts) (D4Y1)
	360 mph (313 knts) (D4Y2)
	357 mph (310 knts) (D4Y3)
	350 mph (304 knts) (D4Y4)
Ceiling:	32,480 ft (9,900 m) (D4Y1)
	35,105 ft (10,700 m) (D4Y2)
	34,450 ft (10,500 m) (D4Y3)
	27,725 ft (8,450 m) (D4Y4)
Range:	2,417 miles (2,100 naut. mls) (D4Y1)
	2,239 miles (1,945 naut. mls) (D4Y2)
	1,796 miles (1,560 naut. mls) (D4Y3)
	1,611 miles (1,400 naut. mls) (D4Y4)
Engines:	One 1,200-hp Aichi AE1A, 12-cylinder, liquid-cooled (D4Y1)
	One 1,400-hp Aichi AE1P, 12-cylinder liquid-cooled (D4Y2)
	One 1,560-hp Mitsubishi MK8P 14-cylinder, liquid-cooled (D4Y3/4)
Armament:	Two fuselage mounted 7.7 mm machine guns
	One flexible rearward firing 7.92 mm machine gun
Crew:	Two

Right: Admiral Isoroku Yamamoto

Below: Admiral Chester Nimitz with Admiral Halsey..

Japanese attack on Pearl Harbor showing the first bombs
hitting American battleships.

Mitsubishi G3M-25s.

Aichi A6M2 preparing to take off to attack Pearl Harbor.

HMS *Hermes* sinking after being bombed.

Aircraft carrier USS *Lexington* burning after being hit by bombs and torpedoes.

Japanese aircraft carrier *Akagi*.

Above left: USS *Lexington*.
Above right: USS *Bunker Hill*.

Japanese aircraft carrier *Kaga*.

Japanese aircraft carrier *Shoho* after being hit by a torpedo from a Douglas TBD.

Japanese aircraft carrier *Hiryu* burning.

Japanese battleship *Haruna* under intense aerial bombardment from American carrier aircraft.

Japanese aircraft carier *Chiyoda* under attack.

Boeing B-29 bombers unloading their bombs on Japan.

P51 Mustangs that escorted
B-29s over Japan.

Boeing B-29 bombers on a
bombing mission.

Kamikaze pilots.

Above left: Kamikaze aircraft about to crash onto the deck of an American aircraft carrier.
Above right: Posed photograph of kamikaze pilots.

Aircraft carrier being repaired after a kamikaze attack.

B-29 *Bockscar* that dropped the atomic bomb on Nagasaki.

The atomic cloud
over Nagasaki.

CHAPTER FIVE
The Pacific War

On the eve of 7 December 1941, diplomatic negotiations between the American government and Japan collapsed. Within hours Admiral Isoroku Yamamoto had sent the order to Vice-Admiral Chuichi Nagumo – commander of the Special Carrier Force, consisting of six aircraft carriers, *Akagi*, *Hiryu*, *Kaga*, *Soryu*, *Zuikaku* and *Shokaku* and a number of other warships all anchored at Hitokappu Bay off the southern part of the Kuril Archipelago – to attack Pearl Harbor. The fleet weighed anchor and at 0600 hours on 7 December 1941, the first of twenty-five Aichi D3A (Val) dive bombers roared off the decks of the carriers, followed closely by forty-nine Nakajima B5N (Kate) carrier attack bombers and forty-three Mitsubishi A6M (Zeke) fighters.

At 0700 hours, the first wave of aircraft led by Cdr Mitsuo Fuchida, consisting of forty-nine Nakajima B5N 'Kate' torpedo bombers, forty-nine 'Kates' carrying 1,760 lb armour piercing bombs, fifty-one Aichi bomber and forty-three Zeke fighters, swept in at low level dropping their torpedoes and bombs into the harbour, and within minutes a number of American battleships were either on fire or sinking. The Zekes that followed the bombers attacked the Hoiler Air Base, Hawaii, strafed the parked aircraft, and attacked the few American fighters that managed to get airborne. One hour and ten minutes later, the second wave, led by Lt-Com. Shimazaki and consisting of fifty-four 'Kate' bombers, eighty 'Vals' and thirty-six Zeke fighters, struck Pearl Harbor again, causing further devastation. The attack had caught the Americans completely by surprise and decimated the United States Pacific Fleet. A total of 353 Japanese aircraft took part in the Hawaiian Operation, but surprisingly only 154 attacked Pearl Harbor; the remaining 199 fighters and bombers were assigned to attack the American airfields and destroy all the aircraft on the ground and any that managed to get into the air, thus marking their air superiority. A total of 188 American aircraft were destroyed during the raid, while the Japanese lost just twenty-nine aircraft.

The returning Japanese aircraft were greeted with adulation from the crews on the aircraft carriers; the returning flight commanders urged Admiral Chuichi Nagumo that it would be a good time to carry out another strike on Pearl Harbor, but he decided that they had done enough and that they had achieved a great victory. Three of the American battleships had been sunk – the *Arizona*, *Utah* and *Oklahoma* – and a large number of the other warships had just been damaged. These were later repaired and rejoined the fleet. Had the Japanese been aware of the position of the aircraft carriers, there is no doubt that they could have attacked and sunk them. The Japanese also hardly touched the huge docking and repair facilities and the massive fuel depot containing over four million barrels of fuel oil. Had they destroyed the fuel depot, the US Pacific Fleet would have had to withdraw to the West Coast and carry out the fight from there. In Japan, Admiral Yamamoto warned the Supreme Headquarters that they had now 'Grabbed a tiger by the tail' and that the war with the West would have to be concluded within two years, otherwise Japan would feel the might of the Western world.

The Supreme Headquarters announced to the public that the American Pacific Fleet had been totally destroyed by a victorious Japanese Navy. The euphoria that followed the attack would have been short-lived had the Japanese realised that three large American aircraft carriers, twenty-one cruisers, fifty-eight destroyers and twenty-two submarines were still active in the Pacific. In addition to this, nearly all the ships lost at Pearl Harbor were nearing the end of their operational lives and were due for replacement. However, this was still a devastating loss to the United States of both men and ships.

The war in the Pacific had begun, and Japanese forces started their attacks on the Philippines, Singapore and the Dutch East Indies. At the same time as the attack on Pearl Harbor took place, Admiral Yamaguchi's force attacked Wake Island. The Americans initially repulsed the attack and inflicted heavy casualties on the Japanese invaders. Because they had no aircraft carriers in the vicinity, the invading troops had no air support, but on 21 December Admiral Yamaguchi's 2nd Carrier Division arrived and launched eighteen Val dive-bombers escorted by eighteen Zeke fighters. Because of heavy cloud over the island, the Japanese were unable to spot the targets and engage any enemy fighters. The following day, with the weather cleared, they again attacked the island, this time inflicting heavy casualties on both aircraft and equipment, including the large guns that had caused so much damage to their ships during the first attempted invasion of the island. The Val dive-bombers and Zeke fighters quickly dealt with the few Wildcat fighters that managed to get off the ground and took control of the air over Wake Island. Within weeks, the Japanese marines had taken control of the island and another foothold in the Pacific had been gained.

The third day of the Pacific War produced another major victory for the Japanese when two of Britain's warships, the battleship *Prince of Wales* and the battlecruiser *Repulse*, were sunk. The two ships had been seen leaving Singapore and had been tracked by the submarine I-56. When the message was received, a formation of nine Mitsubishi G3M 'Nell' bombers, two Mitsubishi C5M 'Babs' reconnaissance aircraft, twenty-six 'Nell' and twenty-six Mitsubishi G4M 'Betty' bombers carrying torpedoes, and thirty-four 'Nell' bombers carrying additional bombs took off to find and destroy the two battleships. As the formation sighted the two battleships and their escorts, they manoeuvred themselves into an attack formation. Surprisingly they encountered no air opposition, and were given free rein to attack. Both ships were sunk within a short period of time with the loss of 840 men, including Admiral Sir Tom Phillips. This was a devastating loss to the British and one that was to take some time to recover from. It was discovered later that the moment the two battleships had been discovered, some of the formation had broken away to attack the nearby Kuantan air base to prevent the Allied fighters taking off.

In the meantime Rear-Admiral Yamaguchi's force, which had participated in the attack on Pearl Harbor, returned to Hiroshima Bay on 23 December to refuel and rearm. Waiting for them was Admiral Yamamoto, to congratulate them on their victory. Two weeks later the task force, under the command of Vice-Admiral Nagumo, left Hiroshima Bay to support the assault operations on Rabaul (New Britain) and Kavieng (New Ireland), and later on New Guinea.

On 15 February, the attacking force made its way through the Dutch East Indies towards the northern coast of Australia. On 19 February a force of 190 bombers and fighters attacked Port Darwin, shooting down eight Curtiss Tomahawk fighter aircraft and destroying another fifteen on the ground. In addition, Val dive-bombers sank two destroyers and a number of other smaller vessels in the harbour. The attack revealed that the Allied aircraft were no match for the Zeke fighters. The resistance met was very limited.

At the same time as the attack on Port Darwin was being made, land-based aircraft from the newly occupied airfields in Southern Borneo were attacking selected targets in the Java Sea. Then, on 27 February, a Japanese reconnaissance plane spotted the American seaplane tender USS *Langley* south of Java Island, accompanied by two destroyers. Nine Mitsubishi G3M 'Nell' bombers escorted by six Zeke fighters were sent to attack the carrier. The attack was a complete success. The bombers hit the *Langley* with five direct hits, causing it serious damage. Later the escorting destroyers took off the crew, and the now-drifting hulk was sunk by torpedoes fired by one of the escort destroyers.

The Japanese were beginning to control almost all the air space in the Pacific as their attacking forces continued to sweep across the Indian and Pacific Oceans. The speed of the advance took even the Japanese by surprise, and the first of the concerns that were to affect the military began to manifest itself. The problem that faced the rapidly advancing armies was that their supply lines were being stretched almost to breaking point, and in the coming years the Allies were able to exploit this by cutting off these lines.

In April, the Allies suffered a severe loss when the British aircraft carrier HMS *Hermes* was sunk off Ceylon after being hit by over forty 250 lb bombs.

THE BATTLE OF THE CORAL SEA

In May 1942 one of the most decisive battles of the war took place, the Battle of the Coral Sea. The battle came about after it was discovered that an American aircraft carrier, accompanied by a number of other warships, was in the Coral Sea. This was at a time when the Japanese planned to invade New Caledonia and Fiji. The island of Tulagi had been captured by Japanese land forces in order to set up a forward base for the proposed invasion of New Caledonia and Fiji. Then, on 6 May, a Kawanishi H6K2 'Mavis' flying boat spotted an enemy force that included at least one aircraft carrier and nine other warships. The following day an intensive air search was made to locate the American aircraft carrier and her escorts. Then came the breakthrough: a Nakajima B5N1 spotted them just 200 miles south of the Japanese aircraft carriers *Shokaku* and *Zuikaku*. Every available aircraft was launched from both carriers, a total of thirty-six 'Val' dive-bombers, twenty-four 'Kate' torpedo bombers and eighteen 'Zeke' fighters. They were to be joined by sixty Mitsubishi G3M 'Nell' bombers and eighteen Zeke fighters. With all the aircraft airborne, a 'Glen' reconnaissance seaplane from the heavy cruiser *Kinugasa* spotted another American aircraft carrier with ten escort ships heading into the area. With all their aircraft committed, the Japanese carriers had to wait the return of their aircraft before refuelling and re-arming them and sending them to attack the other aircraft carrier. As they waited, lookouts aboard the Japanese carrier *Shoho* spotted a large force of almost 100 American bombers and fighters heading their way. Minutes later the *Shoho* was struck by a number of bombs and torpedoes, and was sunk. This was the first Japanese aircraft carrier to be lost.

Admiral Takagi realised that because of the large number of aircraft that had attacked the *Shoho*, he faced at least two large American aircraft carriers. As he waited for his aircraft to return, Takagi calculated that the American carriers were at least 350 miles away, which was beyond the range of any of his aircraft. In addition to this, if he sent some of his aircraft to find and attack the American fleet, they would have to return in the dark. The Zeke fighter escorts were not equipped for night flying, and so would have to be dropped from any mission after dark.

Takagi ordered twelve 'Val' dive-bombers and fifteen 'Kate' torpedo bombers to find and attack the American carriers. As the attack group took off, the weather closed in and after hours of searching it was realised that this was an impossible task and they were running desperately low on fuel. The force headed back to find its own carriers, dropping bombs and torpedoes into the sea in an effort to lighten the aircraft and conserve fuel. Unknown to them, they had passed directly over an American carrier and had been spotted. Suddenly, as if from nowhere, Grumman F4F Wildcats attacked them, resulting in the loss of eight 'Kate' torpedo-bombers and one 'Val' dive-bomber. The remaining aircraft managed to get away and continued back towards their own carriers. Frantically looking for their carrier, they suddenly 'sighted' what they thought was a friendly aircraft carrier and, switching on their landing lights, made their approach. As the first of the aircraft, with its flaps down and speed reduced, came in, the pilot realised that he was approaching an American carrier and quickly raised his flaps and increased the power, then wheeled away from an astonished deck crew.

The remaining pilots quickly realised their mistake and climbed away without a shot being fired at them. Although they were thankful for their lucky escape, they were also angry with themselves for having dropped all their bombs and torpedoes. Only half of the remaining force managed to return; some of those simply ran out of fuel and crashed into the sea, others suffered engine problems. The Japanese were beginning to realise the problems that they faced when operating a strike force in a vast ocean. In an effort to claim some success, on 7 May a 'Glen' reconnaissance seaplane from the *Kamikawa-Maru* spotted an enemy fleet consisting of two battleships, one heavy cruiser and four destroyers 150 miles from Deboyne Island, just off the coast of New Guinea. Admiral Yamada, who was on Rabaul at the time, immediately ordered thirty-three Mitsubishi G3M 'Nell' bombers with an escort of eleven Zekes into the air with orders to attack and destroy the fleet. On their return, pilots claimed to have sunk one battleship, seriously damaged the other battleship and left the heavy cruiser in flames and sinking. It was discovered later that not one of the ships had even been hit, let alone sunk! This was put down to the evasive skill of the ships, and the fact that the crews of the bombers were very inexperienced replacements with no battle experience at all. This immediately flagged up a warning signal that in this first action in which replacement crews had been used, the reserve supply of pilots was woefully inadequate.

In the early hours of the morning of 8 May, in a concentrated effort to find the American fleet, the Japanese carriers *Shokaku* and *Zuikaku* launched a number of reconnaissance aircraft. As dawn broke, thirty-three 'Val' dive-bombers, eighteen 'Kate' torpedo bombers and eighteen 'Zeke' fighters were launched in a wide search pattern to seek the fleet. Some hours later the call came in that the fleet had been sighted by one of the 'Kate' dive-bombers flown by Warrant Officer Kenzo Kanno. He had shadowed the fleet, giving their position to the remainder of the aircraft. Running dangerously low on fuel, he turned to return to his carrier and noticed that the other aircraft that were racing towards him were out of position, and would not spot the enemy ships. Placing himself 'beyond the point of no return', he manoeuvred his aircraft alongside the lead aircraft and led them towards the fleet.

The aircraft spotted the American fleet and attacked. The two carriers, USS *Yorktown* and USS *Lexington*, were the main targets and claims were made that numerous torpedo and bomb hits were made, causing serious damage to both carriers. In fact, the *Lexington* was hit with two torpedoes and two bombs, while the *Yorktown* received one bomb and two near misses. The Japanese lost over half of their entire bomber force, including Warrant Officer Kanno. While this attack was going on, the Japanese carriers *Shokaku* and *Zuikaku* found themselves under attack from a force of more than eighty dive-bombers and fighters. Luck was with the *Zuikaku* when the captain managed to place his carrier beneath a large rain-squall, which prevented the attacking aircraft from pressing home their attack. The *Shokaku* was not so lucky and, among other damage, suffered three major hits to her deck, shattering it, which prevented her releasing and landing aircraft. Despite the damage, she managed to limp back to Japan, where major repairs were carried out. This, of course, had the effect of removing one major aircraft carrier from the scene of action for a considerable time. While the *Shokaku* made her way back to Japan, the *Zuikaku* had the problem of recovering the *Shokaku*'s aircraft, and this meant jettisoning damaged aircraft over the side to make way for them, something the Navy could ill-afford to do. Initially the Battle of the Coral Sea was over, and reports back to Japan claimed a massive victory with the sinking of the American aircraft carriers USS *Lexington* and *Yorktown*. In fact, neither ship was sunk during the battle and only sustained relatively minor damage. The Japanese Navy had lost a large number of experienced pilots and crew members, most of whom were veterans of the Sino-Japanese conflict, and was having to replace them with men who had no battle experience at all.

THE BATTLE OF MIDWAY

Because of the poor results of the Coral Sea battle, the General Staff in Japan postponed indefinitely the invasion of Port Moresby and the surrounding areas. It was considered a minor setback, but a review of their successes, which include the sinking of two enemy aircraft carriers, the USS *Langley* and HMS *Hermes*, seriously damaging another, sinking or damaging at least ten battleships and destroying four heavy cruisers and ten destroyers, all within a six month period, placed the balance of the war firmly in their favour. The losses suffered by Japan during this period were very light in comparison, but the Battle of Midway was to change all this and become a major turning point in the war.

At the beginning of June 1942, the ongoing disagreements between the Japanese Army and the Navy came to a head when the Navy proposed that there should be an invasion of Hawaii and Australia in order to expand Japan's territory as quickly as possible. The Army, on the other hand, wanted to occupy the Aleutian Islands, New Caledonia, Ceylon and the Cocos Islands. The Army argued that to invade Hawaii and Australia would stretch supply lines to breaking point and leave them vulnerable to attack. After much arguing it was decided to go with the Army's plan, and on 4 June, two aircraft carriers, the *Ryujo* and *Shinyo*, launched an air attack against the American base at Dutch Harbor, Unalaska Island in the Aleutians. On the following day, 5 June, a force of four aircraft carriers, the *Kaga, Akagi, Soryu* and *Hiryu*, four battleships, and over sixty other warships arrived 170 miles off the island of Midway.

What Admiral Yamamoto did not know was that the Japanese code had been broken, and that Admiral Nimitz and his staff were well aware of his plans. On 3 June, Yamamoto had placed a line of submarines across the Pacific to report the movements of the Pacific Fleet when it left Pearl Harbor, but by the time the submarines were in position, the American fleet, well aware of the plan, had sailed past and the submarines saw nothing throughout the entire battle. Yamamoto had also been given information saying that both the American carriers in the Coral Sea had been either sunk or so badly damaged that they would be unable to take part in any action. As the American fleet neared Midway, Admiral Fletcher, because of the intercepted messages, had a rough idea where Admiral Nagumo's fleet was. Nagumo, however, had no idea that the American fleet was even at sea.

Just before dawn on 4 June, a total of 108 aircraft, comprising of thirty-six Zekes, thirty-six 'Betty' bombers and thirty-six Type 99 'Kate' dive-bombers, were launched from the Japanese carriers to attack the airfield on Midway. The aircraft on Midway were no match for the Zekes, and almost all the aging Brewster Buffaloes and Grumman Wildcats were shot down. However, the gunners on Midway fared much better and shot down almost one third of the enemy strike force.

As the remnants of the first wave were returning from the raid, a reconnaissance aircraft from one of the escorting destroyers spotted an American aircraft carrier in the distance, heading towards Midway. Some minutes later, one of the 'Betty' bombers returning from the raid spotted a second American aircraft carrier, but failed to notify his commander until he was back aboard the *Soryu*.

Frantic preparations were under way on all the Japanese carriers for a second wave of bombers to be launched against Midway, and the word that there were two large American aircraft carriers close by became a matter of serious concern. The flight commanders had just rearmed their aircraft with high-explosive bombs for the continuing attack on Midway, and with the news that American carriers were within the vicinity it was decided to rearm some of the dive-bombers with torpedoes. This meant that the decks of the carriers were littered with high-explosive bombs and torpedoes as the deck crews worked frantically to replace and rearm the bombers.

Within a few minutes of the warning of the American carriers, forty-one Douglas TBDs suddenly appeared at wave-top height. Japanese Zeke fighters, who were flying a protective cover around the fleet, attacked and shot down thirty-eight of the attackers. With the attack successfully warded off, the crews relaxed, but then suddenly out of the clouds above the carriers appeared a force of eighty-two Douglas SBDs, diving towards them. The Zekes were slow to react and within minutes the *Akagi, Kaga* and *Soryu* were on fire, the bombs and torpedoes on the decks exploding under the intense heat. As the SBDs pulled away, a second wave of American bombers from the aircraft carriers USS

Hornet, *Yorktown* and *Enterprise* took their place, and within minutes three of the four Japanese carriers were lost.

The only survivor was the *Hiryu*, which, although badly damaged, was able to launch forty of her aircraft to attack the USS *Yorktown*, inflicting serious damage. But within minutes of the last of her aircraft being launched, the *Hiryu* was a mass of flames as Douglas SBDs from the *Yorktown* struck back. The loss of her four main aircraft carriers was a devastating blow to the Japanese Navy, who also lost 285 aircraft out of a force of 372, a cruiser and a submarine. The Americans lost one aircraft carrier, the USS *Yorktown*, 150 aircraft out of a force of 307, and one destroyer.

In order to save face, the Supreme Headquarters put out a statement saying that only one of their aircraft carriers had been sunk and one damaged. The landings on two of the Aleutian Islands were successfully carried out, but because of the major losses the planned invasions of New Caledonia, Fiji and Samoa were cancelled.

It is interesting to note that the Army was more concerned by what was happening in Manchuria and China than what was happening in the Pacific. This was mainly because, after the Battle of Midway, the Pacific theatre of war became relatively quiet. There were the odd skirmishes in Burma and the Philippines, but in general between June and July nothing major erupted. The Army then suddenly decided to split its air section into three Air Commands. One third went to Manchuria, one third remained in the Pacific, and the remaining third went to Japan to protect the homeland. The Navy, in effect, was going to have to control the air over the Pacific theatre of war on its own. With this in mind, on 6 July 1942, 200 Japanese marines and 2,000 labourers landed on a small island in the far south of the Pacific Ocean called Guadalcanal, to build an airfield. This tiny island, which, initially, no one had heard of, was later to become the centre of some of the bitterest fighting of the Pacific War. Because of its remoteness, the Japanese Supreme Headquarters did nothing to consolidate it and left but a token force to protect it.

The Japanese Army at this time was convinced that its ally Germany, which was advancing across North Africa, was about to take the Caucasus and would ultimately join forces with the Japanese Army in India. Hitler's promises convinced the Japanese that this was going to happen, but like many other countries that listened to his ramblings, they were disappointed, and in 1944 Japan found herself alone, fighting for her survival against overwhelming odds.

GUADALCANAL

With the Japanese still smarting from their losses at Midway, General Douglas MacArthur decided now was the time for a defensive counter attack and proposed a direct assault on Rabaul, in New Britain. This would have to be a combined land and sea operation, but Admiral Chester Nimitz was reluctant to risk his few remaining aircraft carriers in such an assault. It was decided that the Navy would carry out an assault on the Santa Cruz Islands and the seaplane base at Tulagi, and when secured, MacArthur would carry out an attack on Papua and the Solomons and converge on Rabaul. The attack was imminent when news came through from coast-watchers that the Japanese were building an airstrip on the north coast of Guadalcanal. The island was one of the chains of islands that made up the Solomon Islands, and was the only one suitable for an airbase. The remaining islands were mountainous and jungle-clad.

Strategically, the island of Guadalcanal was priceless, and any planned attack on Papua had to be put on hold until such time the island was in the hands of the Americans. One of the problems that faced the Americans was that there were no maps or charts for the islands, and any decisions had to be made with incomplete data and guesswork. At first, General Vandegrift, the US Marine commander, asked for a postponement so that more information about the island and the reef surrounding it could be obtained, but he was told a seven-day deferment was all that could be allowed.

On 7 August 1942, just one month after the Japanese Marines had taken Guadalcanal, the United States 1st Marine Division landed on the island, supported by a small number of P-39s of the 67th and 339th Squadrons. Within thirty-six hours they had taken the northern part of the island and its airfield, and renamed it Henderson Field. Suddenly, the Japanese Supreme Headquarters realised the importance of losing the island; the Americans now had an air base from which they could control the air in that part of the Pacific. Had the Japanese mounted a swift counter-attack then they might have re-captured the island, but they waited some time before doing so, and by this time the island had been reinforced by both men and aircraft. Over the next six months, despite desperate attempts to recapture the island, the Japanese attacks failed, and it soon became obvious that the Americans were more than sufficiently well equipped, both in men and supplies, and that the Japanese were losing more men every day.

The capture of Guadalcanal by the Americans had taken place just as the Japanese Navy were about to launch an attack on the Allied air base at Rabi on the south-eastern end of New Guinea. The Allies' main air base was at Port Moresby, but neutralizing the Rabi air base would reduce the number of operational airfields. The main Japanese air base in the area was at Rabaul, on the eastern end of New Britain Island, and it was from here that they launched their attacks against Port Moresby. When the Japanese pilots were briefed for the attack on Guadalcanal, not one of them had heard of it, let alone knew where it was. They were to fly one of the longest missions to date – a round trip of over 1,000 miles. Coupled with this information came the news that American aircraft carriers were in the vicinity of Guadalcanal. This meant that they would be facing experienced Navy fighter pilots, and some of the Japanese pilots had only faced Chinese pilots during the Sino-Japanese incident. The force consisted of twenty-seven Mitsubishi Navy Type 1 Attack bomber (G4M), nine Kawasaki Attack bombers (Ki-48) and seventeen Zeke fighters flying escort.

While the land battle for Guadalcanal raged, at sea fighting also raged when the US fleet appeared. The Japanese sent four of their aircraft carriers, the *Zuikaku, Shokaku, Zuibo* and *Ryujo*, into the fray. On 25 August, the aircraft carrier *Ryujo* was sunk by dive-bombers from the USS *Enterprise*, but the *Enterprise* also suffered some damage from 'Val' dive-bombers from the carrier *Shokaku* and had to withdraw for repairs. While the Japanese Navy concentrated its attack on the American fleet, they failed to notice twenty-five American supply ships slip in and re-supply the Marines with everything that they required. The Japanese did have one success: they sank the aircraft carrier USS *Wasp*. The submarine I-19 sank the carrier as it made its way back to join the rest of the US Fleet.

It soon became obvious to the Japanese Supreme Headquarters that the battle for Guadalcanal was a lost cause; the Americans maintained air superiority over the islands. The Japanese were unable to maintain supplies to their beleaguered troops, and in February 1943 they abandoned the island, leaving behind 24,600 dead Marines and infantry. This was a bitter pill for the Japanese to swallow and the news was not well received back in Japan, as it appeared that the men had died for nothing.

The constant missions were beginning to take their toll on the Japanese pilots as they became increasingly tired and weary. Also, the number of experienced pilots and crews being lost was becoming a cause for concern among the senior members of the air arms. The Allied forces were growing stronger by the day, and the massive manufacturing plants in the United States were producing more and more aircraft, especially long-range bombers like the B-17 Flying Fortresses and Consolidated Liberators and fighter aircraft like the P-38 Lightning, Grumman Hellcat and Avenger. The Japanese aircraft manufacturers were barely managing to keep pace with the demand for existing models of aircraft, which meant that there were no new types or models to replace them, whereas the Americans were bringing out faster and more powerful aircraft as the war progressed.

With the success of Guadalcanal, the Americans pushed to take the Solomon Islands. The Japanese, still smarting from their forced retreat from Guadalcanal and their defeats in New Guinea, sent a force of 7,000 troops from Rabaul in an effort to regain parts of

New Guinea. On 1 March 1943, escorted by eight destroyers, the infantry division was embarked on seven transports and one converted collier. Undetected by the Allies, the convoy entered the Bismarck Sea during bad weather, but on 2 March, a reconnaissance aircraft of the RAAF spotted the convoy steaming towards New Guinea. A force of RAAF and USAAF bombers and fighters from Papua New Guinea attacked the convoy over the next two days. All the transports and four of the destroyers were sunk for the loss of two bombers and three fighters. The success of this mission highlighted the capabilities of the Allied air forces to attack enemy forces far out at sea without the support of Naval ships.

The Japanese continued to expand and reinforce their bases in the Solomon Islands, but Yamamoto regarded these as defensive measures, and he wanted to take the fight to the Allies. At the beginning of April 1943, he launched I-go Operation. This was an air attack, consisting of 170 aircraft from his four aircraft carriers and 200 aircraft from shore-based airfields, on Allied shipping at Guadalcanal, Tulagai, Port Moresby and Oro Bay, near Buna. Returning aircrews announced tremendous successes, claiming the destruction of various warships belonging to the United States and Australia. In fact, they sank one US destroyer, one tanker, an Australian minesweeper and a New Zealand anti-submarine trawler, and damaged a small number of merchant ships. Over fifty aircraft and their crews were lost, a loss that the Japanese Navy could ill-afford to take. In fact, the Japanese had lost an estimated 2,500 aircraft and crews in their attempt to recapture Guadalcanal and the Solomon Islands. As far as the air war was concerned in the Pacific, the Allies were now in control and the Japanese were relegated to fighting a rearguard action. It was now up to the Allied navies to take control of the seas.

Then, on 18 April 1943, on what was to be a surprise inspection of the front line, Admiral Yamamoto's bomber was shot down by P-38 Lightnings of the USAAF 13th Air Force. Yamamoto was a stickler for punctuality, and had planned his visit down to the last minute. Notification of his visit was sent to the various bases at the very last minute, but the Allies were monitoring the radio traffic between the bases. Once they knew of his proposed visit, plans were put into action to intercept him and his escort, knowing full well where he was going to be at any given time. As the formation of two Mitsubishi Army Type 97 'Betty' (Ki-21) heavy bombers escorted by nine Zeke fighters approached the south-western tip of Bougainville Island, dead on time, twenty-four P-38 Lightnings pounced. Yamamoto's aircraft, the other bomber and a number of the escorting fighters were shot down. There were no survivors from Yamamoto's bomber, but the other bomber managed to make a crash landing in the sea; only two survived, Vice-Admiral Matome Ugaki and Rear-Admiral Kitamura. The death of Yamamoto was a massive blow to the Japanese, as he was regarded as the architect of the war in the Pacific, and the one to whom all the other admirals looked towards for guidance.

At sea, the battles between the two sides seemed to be going the way of the Japanese, but the loss of their ships was causing the Japanese more problems, since their replacements were almost non-existent. The Americans, on the other hand, seemed to have no problem replacing their destroyers and cruisers.

SOLOMON ISLANDS

One of the turning points in the war at sea occurred on 5 November 1943, when Vice-Admiral Takeo Kurita led a formation of six heavy cruisers in an attempt to maintain control of the seas around the Solomon Islands. TBF Avengers and Douglas TBDs from the aircraft carriers USS *Saratoga* and *Princeton* intercepted the formation. Five of the six cruisers were badly damaged, forcing the formation to retreat to their base in Truk. Control of the seas around the Solomon Islands was now firmly in the hands of the Allies.

On 11 November a carrier strike force, consisting of the aircraft carriers USS *Bunker Hill*, *Essex* and *Independence* escorted by ten destroyers, attacked Japanese ships in Rabaul Harbour, damaging two destroyers and a cruiser. Seeing an opportunity to possibly regain control of the Solomons, Admiral Koga, who had replaced Yamamoto, ordered a massive air strike against the strike force: twenty-seven Type 99 D3A 'Val' Carrier Bombers and fourteen B5N 'Kate' Carrier Attack Bombers escorted by sixty-seven Zekes. The massive formation thundered towards the American strike force, only to be met by F6F Wildcat and F4U Corsair fighters from the carriers. The inexperienced Japanese pilots were no match for the battle-hardened American fighter pilots and twenty-four 'Vals', twelve 'Kates' and thirty-two 'Zekes' were shot down. The American ships suffered no damage, and only a small number of American aircraft were damaged.

The beginning of 1944 heralded the start of a massive defensive campaign by the Japanese, as day after day they were subjected to an increasing number of bombing raids. Control of the seas and the air were now in the hands of the Allies and the pressure on the Japanese was relentless. This was highlighted when the American invasion of the Marianas started, with the assembly of 535 ships carrying or escorting over 129,000 troops, who were able to reach their destination without being attacked.

The morning of 13 June saw the island of Saipan subjected to a massive bombardment from US battleships. The bombardment lasted for two days, and on the morning of 15 June 1944 the Japanese garrison on Saipan were shocked to see that a huge armada of enemy ships had arrived offshore overnight. Within minutes of the sun rising, Grumman F6F Hellcats, accompanied by a massive bombardment of naval guns, were attacking the Japanese positions. In less than an hour, some 8,000 US Marines were ashore and consolidating their positions, and by the end of the day a further 20,000 Marines had arrived. All through that first night the Japanese soldiers carried out screaming *banzai* charges, but with little or no effect.

At sea, the American Fleet awaited the arrival of the Japanese Fleet, in the hope that they could once and for all annihilate it. While the bombardment of Saipan was being carried out, the First Mobile Fleet, commanded by Vice-Admiral Jisaburo Ozawa, sailed from Tawitawi to join up with Rear-Admiral Ugaki's task force. The fleet was split into three groups; the first group, under the command of Vice-Admiral Takeo Kurita, consisted of three light aircraft carriers, *Chitose*, *Chiyoda* and *Zuiho*, carrying eighty-eight aircraft, escorted by four battleships, *Yamato*, *Musashi*, *Haruna* and *Kongo*, four heavy cruisers and six destroyers. The second group, under the command of Vice-Admiral Ozawa, consisted of three aircraft carriers, *Taiho*, *Shokaku* and *Zuikaku*, carrying a total of 207 aircraft and escorted by a number of cruisers and destroyers. The third group, under the command of Rear-Admiral Takaji Joshima, had three aircraft carriers, *Junyo*, *Hiyo* and *Ryuho*, carrying 135 aircraft, the battleship *Nagato*, one cruiser and four destroyers. The Japanese fleet intended to aid the beleaguered troops on Saipan, but between them and Saipan lay the American Task Force 58 (TF58), commanded by Rear-Admiral Raymond Spruance, one of America's top admirals and one capable of operating under enormous stress. Under his command he had fifteen aircraft carriers, carrying a total of 965 aircraft, seven battleships, twenty-one cruisers and sixty-nine destroyers. This vast armada covered an area thirty-five miles wide by twenty-five miles long and encompassed an area of 700 square miles. Despite his superior force, Spruance did not make full use of his air reconnaissance aircraft, while, on the other hand, Ozawa used all of his air-search aircraft to their fullest extent.

On the morning of 18 June, Vice-Admiral Ozawa received the information that two of his reconnaissance aircraft had sighted the American TF. In the meantime, Spruance had received a message from Pearl Harbor that the Japanese fleet was 350 miles south-west of him. One the morning of 19 June, aircraft from the American carriers attacked targets on Guam, Rota, Tinian and Saipan, and also intercepted reinforcements coming in from Truk. This set back Ozawa's plans considerably, as he had hoped that when he attacked the American task force he would have been supported by aircraft from these bases.

Early the following morning, Ozawa launched a number of reconnaissance aircraft in an effort to pinpoint the American task force. Almost all of the aircraft were shot down, but not before they had found and radioed back the exact position of the task force.

Within hours, forty-five Zekes carrying bombs and eight Nakajima B6N 'Jill' carrier attack bombers armed with torpedoes and escorted by sixteen Zeke fighters were on their way to attack the American Task Force. Around 150 miles from their target, they were detected by radar. The carriers immediately launched their F6F Hellcat fighters, giving them sufficient space and height to intercept the Japanese aircraft before they even got near the task force.

The Japanese pilots were not the battle-hardened ones that had been in action during the early part of the war, but novices up against battle-hardened US Navy pilots who could smell success with every mission. Of the sixty-nine aircraft that took part in the raid, only twenty-four survived and most of them were damaged. Admiral Ozawa launched a second strike within hours – an even bigger one than before. It consisted of fifty-three Kugisho D4Y1 'Judy' carrier bombers and twenty-seven Nakajima B6N 'Jill' carrier attack bombers, escorted by forty-eight Zeke fighters. As the Japanese carrier *Taiho* turned into the wind to launch her aircraft, she was hit by a salvo of six torpedoes from a submarine, the USS *Albacore*. Despite the strike, all her aircraft were launched, but of the 137 aircraft launched, ninety-seven of them were destroyed. Throughout the day American aircraft patrolled the task force, but the initial attack in which the majority of the Japanese aircraft were shot down became known as 'the Great Marianas Turkey Shoot'. Admiral Ozawa had lost two aircraft carriers and 346 aircraft and their pilots.

During this aerial battle, none of the task force aircraft had even sighted the Japanese fleet. This was to change when, on 20 June, Rear-Admiral Marc A. Mitscher, who commanded the aircraft carriers, launched an all-out strike against the Japanese fleet. At 4 p.m., a force of eighty-five F6F Hellcats, seventy-seven Douglas Helldivers and fifty-four TBF Avengers was launched in a staggering ten minutes. The Japanese aircraft carrier *Hiyo* was sunk, the *Chiyoda* was badly damaged, and two tankers were sunk. The Japanese Combat Air Patrol (CAP) put up a strong resistance, but they were no match for the Americans. Admiral Ozawa conceded defeat and turned what remained of his fleet for home. He had suffered the loss of 476 aircraft and 445 aircrew while the Americans had lost a total of 130 aircraft.

With no support from the Navy, the resistance on the island of Saipan collapsed and on 9 July 1944 the Americans took control of the island. With defeat staring them in the face, Admiral Nagumo, and his Army counterpart General Saito committed suicide on the island. The vast majority of the Japanese civilian population, having been told that the Americans would subject them to unspeakable things, committed mass suicide by throwing themselves off the cliff at Marpi Point on the north end of the island.

LEYTE GULF

The next major battle, which in reality consisted of four major engagements, and the one that was to put the final nail in the coffin of the Japanese Navy, was the Battle of Leyte Gulf. This battle was one of the most complex naval engagements of the Pacific War. There were problems from the start between the Navy and the Army. The Army, under the command of General Douglas MacArthur, wanted to liberate the Philippines, while Admiral Chester Nimitz had been ordered to take the Marianas and then the Palau Islands. Roosevelt should have made one of the commanders the supreme commander, but for some unknown reason failed to do so, thereby causing a situation where no one person had overall control in the Pacific theatre of war. The preparations for these offensives were to prepare the way for an all-out attack on Japan. MacArthur's troops were already in position at Sansapor, which was on the northern tip on New Guinea, while Nimitz and his fleet were in the Marianas. MacArthur's plan was to take the island of Morotai and the Talaud Islands. He would then build airstrips, so that when he next moved he would have substantial air cover for his ground troops. Nimitz's intention was to carry out an assault against the Paulus Islands and then take the island of Yap.

Task Force 38, under the command of Vice-Admiral Halsey, carried out a number of 'softening-up' strikes against both Paulus and Yap. The opposition was surprisingly light, the Japanese losing over 200 aircraft and a number of small ships against just eight aircraft from the task force. It was decided that an early attack on Leyte in the Philippines would be a much better target. In preparation for the attack, on 15 September 1944, after the usual bombardment from both the sea and air, the 1st US Marine Division went ashore on the island of Peleliu. It soon became obvious why the opposition was so light in Paulus and Yap: the Japanese had reinforced the ground troops at Peleliu. There were over 10,000 troops ensconced in the natural caves that were an integral part of the coastline, making it almost impossible to flush them out. On 20 October, MacArthur's troops joined forces with the 1st US Marine Division and took the fight to the Japanese. The fighting was as bitter as any that had been experienced in the Pacific, and it wasn't until March 1945 that the last of the resistance was finally broken. Adjoining the island of Peleliu was the atoll of Ulithi, which the Japanese had abandoned some months earlier and which had a superb natural harbour. The Third Fleet took advantage of this and made it its main advance base.

With Peleiu now in Allied hands, MacArthur's troops landed on the island of Morotai, one of the most northerly islands of New Guinea. The Allies continued to make progress, and by the beginning of October 1944 had taken control of almost all the islands from the Moluccas to the Marianas. This placed them in a very strong position for their assault on the Philippines, and the date of 20 October was set for the landings on Leyte.

On 10 October 1944, fifteen aircraft carriers of TF38 arrived off the Philippines and began a massive air bombardment in support of the forthcoming Leyte assault. So widespread and heavy were the attacks by over 1,000 aircraft that the Japanese thought that it was the main assault. Japanese aircraft attacked the task force, hitting the heavy cruiser USS *Canberra* and the light cruiser USS *Houston* with torpedoes. The young, inexperienced Japanese pilots returned to their base claiming to have sunk eleven of the aircraft carriers and two battleships. Tokyo Radio's 'Tokyo Rose' broadcast this, and so, convinced that it was true, Vice-Admiral Shima, with two of his heavy cruisers *Nachi* and *Ashigara*, accompanied by a destroyer escort, set off to 'mop up' the remaining part of the task force. As he closed on the task force, he realised that he was being drawn into a trap and that none of the carriers had been touched, let alone sunk. Discretion being the better part of valour, he retreated before being sighted.

The Japanese Navy's remaining main battleships and heavy cruisers were based in Singapore, mainly because it was close to their fuel supplies. The remaining aircraft carriers were all in Japan, frantically attempting to train new air groups to replace the ones lost in the Philippine Sea battles. Control over the Philippines was vital to both sides. From the Japanese perspective, if they lost control they would be cut off from their fuel supplies in Malaya and the Dutch East Indies; from the Allied perspective they would have an enemy rapidly running out of fuel, and this could shorten the war considerably.

In the early hours of 20 October, the first American troops went ashore at Leyte; the recapture of the Philippines had begun. Admiral Takeo Kurita set sail from Singapore with the First Striking Force; they were to engage the American task forces in a running battle for two days. With his force dwindling rapidly as TBM Avengers launched torpedo after torpedo into his ships and the American battleships and heavy cruisers pounded them relentlessly, he withdrew. He had lost one fleet carrier, three light carriers, three battleships, six heavy cruisers, four light cruisers, nine destroyers and numerous aircraft and crews. The Japanese Navy was almost decimated.

By the beginning of 1944 it became obvious to many Japanese that the war was lost, and that the invasion of Japan was imminent. In a last, desperate attempt to ward off any invasion by the Americans, Vice-Admiral Onishi Takijiro created the *Tokkotai* (special attack force). This operation – using aircraft, gliders and submarines carrying large amounts of explosives – was ordered to attack American ships, but without any chance of surviving or returning to base in the event of surviving the mission.

When the *Tokkotai* was first mooted to students at the military academy, not one of them stepped forward to volunteer, knowing full well that it would be a self-imposed death sentence. The majority of those chosen came from the university student population, and the government undertook to shorten the courses so that they could be conscripted into the military. Once in the harsh routine of the military, they were subjected to bullying and brutal corporal punishment. If a student refused, he could be publicly humiliated, subjected to peer pressure, sent to the southern front (where they would almost certainly be killed), or in some extreme cases just taken outside and shot. A small number who did openly refuse found themselves to have 'volunteered' in any case, and were among the first to find themselves in training for the one-way missions.

The stories of 'kamikaze' pilots rejoicing in being chosen to carry out these missions to die for the emperor were mostly a fallacy. There were, of course, the odd pilots who did it voluntarily and with pride, but these were in the minority. Propaganda photographs and newsreel pictures shown to the Japanese people showed groups of smiling young, almost boyish pilots about to board their aircraft to give their lives for their emperor. This was far from the truth, as one Japanese airman, Kasuga Takeo, who looked after the *tokkotai* pilots during their training witnessed. He remembers seeing them the night before they left, drinking sake until they were roaring drunk and smashing up the barracks in rage at the military hierarchy. Some just sobbed aloud as they wrote their wills and letters to their loved ones, crying, 'I do not want to die, I do not want to die.' All but a few said that they were not committing suicide for their emperor, but were being murdered by their government. The creed to which the Japanese armed forces adhered was not 'to kill for your country', but 'to die for your country'.

In Japan, the truth about what was happening in the Pacific to their armed forces was becoming more and more difficult to keep from the public. For years they had been told that the Army and Navy were winning battle after battle, and that the Allies were retreating by the day. Such was the desperation that on 15 October 1944, Rear-Admiral Masafumi Arima, commander of the 16th Air Flotilla, armed his Kugisho D4Y bomber with a 500 lb bomb and deliberately crashed it on to the deck of the aircraft carrier USS *Franklin*, killing three crew members and wounding twelve others, but doing very little damage. This was a sign of things to come.

The first of the kamikaze attacks took place on 21 October, when six pilots from the 201st Air Group, led by Lt Yukio Seti, attacked the heavy cruiser HMAS *Australia* with the intention of crashing their aircraft on to the ship. Fortunately only one of the aircraft managed to hit the ship, causing considerable damage. The remaining five aircraft were shot down. Four days later, the escort carrier USS *St Lo* was not so lucky. Suicide bombers attacked her and another escort carrier, and the *St Lo* was sunk. From then on until the end of the war, almost every day either a group or a single suicide aircraft attacked Allied ships.

At sea, the Allies were building larger and larger fleets of ships in preparation for the invasion of Japan, but first they wanted to take the islands of Iwo Jima and Okinawa. Fighters and bombers from the aircraft carriers carried out ferocious attacks on the Japanese airfields on the islands.

IWO JIMA

On 19 February 1945, the attack on the island of Iwo Jima commenced. This was to be one of the bloodiest battles of the Pacific War, in which 5,931 US Marines were killed, a further 17,272 were wounded, and over 21,000 Japanese troops were killed. The island was just 4½ miles long by 2½ miles wide, at its widest point. Attacks on the island by air and bombardment from the sea had started back in August 1944. Throughout that year and into the next, attacks continued in a 'softening-up' process; this included a day-and-night bombing campaign for the two weeks prior to the invasion. The Japanese were so well dug in that the attacks had little effect and the US Marines, when they landed, were met with a hail of gunfire from almost impregnable positions.

Inch by inch, the American troops made their way inland and for the next twenty-five days fought the Japanese relentlessly. On 16 March, the island of Iwo Jima had been secured at a cost of almost 6,000 US Marines and Navy personnel killed and 17,200 wounded, while the Japanese lost almost 22,000 men. The Battle for Iwo Jima became a legend in the history of the US Marine Corps.

With the island secured and the airfields now operational, B-29 Superfortresses were now able to operate, accompanied by long-range P-51 Mustangs. The focus now shifted to the island of Okinawa, which was possibly the most important of all the islands close to the Japanese mainland.

OKINAWA

On 1 April, after five days of continuous bombing, the invasion of Okinawa began. Unknown to the Americans, the Japanese garrison on the island consisted of almost 100,000 men, concentrated in the area around Shuri, the ancient capital of Okinawa. There were also four all-weather airfields on the island, plus another on the nearby island of Ie Shima. In an effort to boost the defence of Okinawa, the Japanese had 6,000 aircraft assigned to the defence of the island, based in Japan. Of these, 4,000 were designated as kamikaze aircraft.

Because the Japanese did not have sufficient troops to defend the outlying islands around Okinawa, the Allies easily took Ie Shima, killing over 3,000 troops, and they established an anchorage there for the fleet tenders and repair ships. By the end of the first day of the invasion of Okinawa, the Americans had landed 50,000 troops, and by the following morning they had started to make their way inland. With an armada of some 1,457 Allied ships in the surrounding waters, the island of Okinawa was cut off from Japan and as such could not be reinforced with men or supplies.

In response to this, the Japanese launched a series of kamikaze attacks, which lasted for two days. Over 600 aircraft were used; 380 were shot down, but the remaining aircraft attacked the fleet. Two American destroyers were sunk, also two ammunition ships. The aircraft carrier USS *Hancock* suffered damage to her wooden deck, which necessitated her having to withdraw for repairs. British aircraft carrier decks were made of steel and in the event of one being hit by a Kamikaze, all that was usually required was a debris-clearing team, and possibly a welder to repair any metal damage. Later American carriers were fitted with steel decks.

The battle for Okinawa raged for almost three months before it was deemed to be secure, at a cost over 7,000 American soldiers killed and over 31,000 wounded. The US Navy lost almost 5,000 men and had 4,800 wounded, while the Royal Navy had eighty-five killed and eighty-three wounded.

With the airfields on Okinawa operational, B-29 Superfortresses escorted by P-51 long-range Mustangs started their relentless bombing campaign on the Japanese mainland. Because of the desperate fuel shortages, many of Japan's larger warships were unable to put to sea, so they were turned into floating anti-aircraft batteries. But they were unable to make any impression on the swarms of bombers that continually pounded Japan's cities.

At the beginning of 1945, the B-29 Superfortresses had begun bombing the aircraft and engine manufacturing plants on the Japanese mainland and within months had brought the aviation industry in Japan to a standstill. The unescorted B-29 bombers did come under attack from home-based fighter aircraft, but the casualties were minimal. When, in May 1945, the B-29s started to carry out bombing raids escorted by long-range P-51 Mustangs, the token Japanese resistance petered out. These raids, together with raids on Japanese Army air bases by US Navy carrier fighters, meant that the bombing of the major cities with incendiary bombs was almost unopposed. The firestorms killed hundreds of thousands of Japanese and made millions homeless.

The Battle of Okinawa in April was the last defiant gesture by the Japanese military, and in August 1945 Japan surrendered, utterly defeated. All remaining Japanese aircraft were grounded and disabled. With propeller aircraft, simply removing the propeller achieved this. The war was over and the Japanese aviation industry never recovered.

The dropping of the atomic bombs on Hiroshima and Nagasaki caused Emperor Hirohito and his government to agree to unconditional surrender. But even on the day of the official surrender, a couple of kamikaze pilots made a last defiant gesture and attempted to attack an American aircraft carrier. The two aircraft were shot down before getting anywhere near the fleet.

At the beginning of the Pacific War the Japanese aircraft manufacturers were having trouble keeping up with demand, and so no new types of aircraft emerged after 1941. There were variations of the existing models, but they alluded in the main to slightly more powerful engines and experiments with weaponry. In the early years Japanese aviation had made gigantic strides, but after entering into war with the Western world, it suffered a series of major setback from which it never recovered.

Appendix – Allied Code Names

A list of Allied code names for Japanese Second World War aircraft. The general rules were:

1. Fighters and reconnaissance seaplanes had boys' names
2. Bombers, dive-bombers, torpedo-bombers, seaplanes and reconnaissance aircraft were given girls' names
3. Transport aircraft were given girls' names beginning with 'T'
4. Trainers were given the names of trees
5. Gliders were given the names of birds

Alf	Kawanishi Navy Type 94 Reconnaissance Seaplane (E7K)
Ann	Mitsubishi Army Type 97 Light Bomber (Ki.30)
Babs	Mitsubishi Army Type 97 Reconnaissance Plane (Ki-15)
Babs	Mitsubishi Navy Type 98 Reconnaissance Plane (C5M)
Baka	Kugisho Navy Special Attack Aircraft (Ohka)
Belle	Kawanishi Navy Type 90-2 Flying Boat (H3K1)
Betty	Mitsubishi Navy Type 1 Attack Bomber (G4M)
Betty	Mitsubishi Navy Type 1 Escort Fighter (G6M1)
Betty	Mitsubishi Navy Type 1 Bomber Trainer (G6M1-K)
Betty	Mitsubishi Navy Type 1 Transport (G6M1-L2)
Bob	Aichi Navy Type 97 Reconnaissance Seaplane
Cedar	Tachikawa Army Type 95-3 Primary Trainer (Ki-17)
Cherry	Kugisho Navy Type 99 Flying Boat (H5Y)
Claude	Mitsubishi Type 96 Carrier Fighter (A5M)
Cypress	Kyushu Navy Type 2 Primary Trainer Momiji (K9W1)
Cypress	Kokusai Army Type 4 Primary Trainer (Ki-86)
Dave	Nakajima Navy Type 95 Reconnaissance Seaplane (E8N1)
Dinah	Mitsubishi Army Type 100 Reconnaissance Plane (Ki-46)
Dinah	Mitsubishi Army Type 100 Tactical Pilot Trainer (Ki-46 IIKAI)
Dinah	Mitsubishi Army Type 100 Air Defence Fighter (Ki-46 IIIKAI)
Emily	Kawanishi Navy Type 2 Flying Boat (H8K)
Emily	Kawanishi Navy Transport Flying Boat Seiku (H8K2-L)
Frances	Kugisho Navy Bomber Ginga (P1Y)
Frances	Kugisho Navy Night Fighter Byakko (P1Y1-S)
Frances	Kugisho Navy Night Fighter Kyokko (P1Y2-S)
Frank	Nakajima Army Type 4 Fighter Hayate (Ki-84)

George	Kawanishi Navy Interceptor Fighter Shiden (N1K1-J)
Glen	Kugisho Navy Type O Small Reconnaissance Seaplane (E14Y)
Grace	Aichi Navy Carrier Attack Bomber Ryusei (B7A1)

Hamp	Mitsubishi Navy Type O Carrier Fighter Model 32 (A6M3)
Hank	Aichi Navy Type 96 Reconnaissance Seaplane (E10A1)
Helen	Nakajima Army Type 100 Heavy Bomber Donryu (Ki-49)
Hickory	Tachikawa Army Type 1 Advanced Trainer (Ki-54a)
Hickory	Tachikawa Army Type 1 Operations Trainer (Ki-54b)
Hickory	Tachikawa Army Type 1 Transport (Ki-54c)

Ida	Tachikawa Army Type 98 Direct Co-operation Plane (Ki-36)
Ida	Tachikawa Army Type 99 Advanced Trainer (Ki-55)
Irving	Nakajima Navy Type 2 Land-Based Reconnaissance Plane (J1N1)
Irving	Nakajima Navy Type 2 Night Fighter Gekko (J1N1-S)

Jack	Mitsubishi Navy Interceptor Fighter Raiden (J2M)
Jake	Aichi Navy Type O Reconnaissance Seaplane (E13A)
Jane	Mitsubishi Army Type 97 Heavy Bomber (Ki-21)
Jean	Kugisho Navy Type 96 Carrier Attack Bomber (B4Y)
Jill	Nakajima Navy Carrier Attack Bomber Tenzan (B6N)
Judy	Kugisho Navy Type 2 Carrier Reconnaissance Plane (D4Y1-C)
Judy	Kugisho Navy Carrier Bomber Suisei (D4Y)

| Kate | Nakajima Navy Type 97 Carrier Attack Bomber (B5N) |

Laura	Aichi Navy Type 98 Reconnaissance Seaplane (E11A)
Lily	Kawasaki Army Type 99 Twin-engined Light Bomber (Ki-48)
Lorna	Kyushu Navy Patrol Plane Tokai (Q1W)
Louise	Mitsubishi Army Type 93-2 Twin-engined Light Bomber (Ki-2-II)

Mary	Kawasaki Army Type 98 Light Bomber (Ki-32)
Mavis	Kawanishi Navy Type 97 Flying Boat (H6K)
Myrt	Nakajima Navy Carrier Reconnaissance Plane – Saiun (C6N)

Nate	Nakajima Army Type 97 Fighter (Ki-27)
Nell	Mitsubishi Navy Type 96 Attack Bomber (G3M)
Nick	Kawasaki Army Type 2 Two-seat Fighter – Toryu (Ki45KAI)
Norm	Kawanishi Navy Type High-speed Reconnaissance Seaplane – Shiun (E15K)

| Oak | Kyushu Navy Type 2 Intermediate Trainer (K10W) |
| Oscar | Nakajima Army Type 1 Fighter – Hayabusa (Ki-43) |

Paul	Aichi Navy Reconnaissance Seaplane – Zuiun (E16A)
Peggy	Mitsubishi Army Type 4 Heavy Bomber – Hiryu (Ki-67)
Peggy	Mitsubishi Army Type 4 Special Attack Plane (Ki-67-IKAI)
Perry	Kawasaki Army Type 95 Fighter (Ki-10)
Pete	Mitsubishi Navy Type 0 Observation Seaplane (F1M)
Pine	Mitsubishi Navy Type 90 Operations Trainer (K3M)

Randy	Kawasaki Army Type 4 Assault Plane (Ki-102b)
Randy	Kawasaki Army Night Fighter (Ki-102c)
Rex	Kawanishi Navy Fighter Seaplane – Kyofu (N1K1)
Rufe	Nakajima Navy Type 2 Fighter Seaplane (A6M2-N)

Sally   Mitsubishi Army Type 97 Heavy Bomber (Ki-21)
Slim   Watanabe Navy Type 96 Small Reconnaissance Seaplane (E9W1)
Sonia   Mitsubishi Army Type 99 Assault Plane (Ki-51)
Spruce   Tachikawa Army Type 95-1 Intermediate Trainer (Ki-9)
Stella   Kokusai Army Type 3 Command Liaison Plane (Ki-76)
Susie   Aichi Navy Type 96 Carrier Bomber (D1A2)

Thalia   Kawasaki Army Type 1 Freight Transport (Ki-56)
Theresa   Kokusai Army Type 1 Transport (Ki-59)
Thora   Nakajima Army Type 97 Transport (Ki-34)
Thora   Nakajima Army Type 97 Transport (Ki-34)
Thora   Nakajima Navy Type 97 Transport (L1N)
Tina   Mitsubishi Navy 96 Transport (L3Y)
Tojo   Nakajima Army Type 2 Single-seat Fighter – Shoki (Ki-44)
Tony   Kawasaki Army Type 3 Fighter – Hien (Ki-61)
Topsy   Mitsubishi Army Type 100 Transport (Ki-57)
Topsy   Mitsubishi Navy Type 0 Transport (L4M1)

Val   Aichi Navy Type 99 Carrier Bomber (D3A)

Willow   Kugisho Navy Type 93 Intermediate Trainer (K5Y)

Zeke   Mitsubishi Navy Type 0 Carrier Fighter (A6M)
Zero   Common name for the Zeke

Bibliography

Francillon, René J., *Japanese Aircraft of the Pacific War*, Putnam 1970

Mikesh, Robert C. & Abe, Shorzoe, *Japanese Aircraft 1910-1941*, Putnam 1990

Mikesh, Robert C., *Japanese Aircraft Code Names*, Schiffer 1993

Mikesh, Robert C., *Zero*, MBI 1994

Sekigawa, Eiichiro, *Japanese Military Aviation*, Ian Allan 1974

Treadwell, Terry C., *Strike From Beneath the Sea*, Tempus 2000

Also abailable from Amberley Publishing

The Spitfire Manual

Edited by Dilip Sarkar

How to fly the legendary fighter plane in combat using the manuals and instructions supplied by the RAF during the Second World War. A fabulous slice of nostalgia, authentic period publications reproduced in facsimile accompanied by contextualising commentary from an expert of the Battle of Britain.

£9.99 Paperback

288 pages, 40 black and white illustrations

978-1-84868-436-2

Also abailable from Amberley Publishing

The Battle of Britain
MINISTRY OF INFORMATION

First published in 1941, *The Battle of Britain* was a propaganda booklet issued by the Ministry of Information to capitalise on the success of the RAF in defeating the Luftwaffe. An amazing period piece, hundreds of thousands of copies were printed and sold for 6d and it became one of the year's best selling books. It is the first book to embed in the public imagination the heroics of 'The Few'.

£4.99 Paperback
36 pages, 25 black and white illustrations
978-1-4456-0048-2

Available from all good bookshops or to order direct
Please call **01285 760 030**
www.amberleybooks.com

Also abailable from Amberley Publishing

German and Austo-Hungarian Aircraft Manufacturers 1908-1918

Terry C. Treadwell

Much has been written about the British aircraft of the First World War, but little has surfaced about the aircraft of the Axis powers, Germany and Austria. Here, Terry C. Treadwell tells the story of the aircraft from companies such as Fokker, builder of the famous triplane, as flown by Baron von Richthofen's Flying Circus, AEG, Albatross, Junkers and Hansa. From reconnaissance aircraft to state-of-the-art bombers that could reach London, this is the definitive guide to aircraft of the Axis powers during the First World War.

£17.99 Paperback

256 pages, 300 black and white illustrations

978-1-4456-0102-1

Available from all good bookshops or to order direct

Please call **01285 760 030**

www.amberleybooks.com

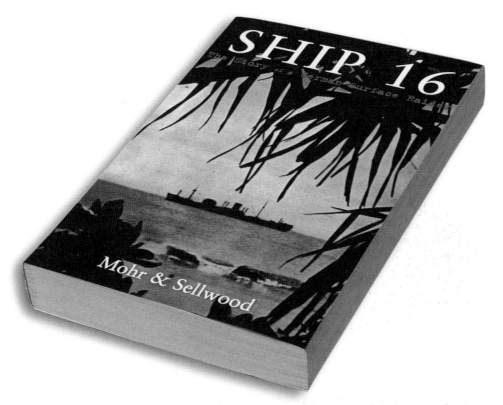

Also abailable from Amberley Publishing

Finest of the Few
The Story of Battle of Britain Fighter Pilot John Simpson

Hector Bolitho

Written by 43 Squadron's intelligence officer, Hector Bolitho, Finest of the Few revolves around Bolitho's friend, fighter ace John W.C. Simpson, who shot down 13 German aircraf during the Battle of Britain. The book was written in 1941 and was based on John Simpson's Combat Reports, his personal letters and papers together with Hector's own recollections of the heady days of the summer of 1940.

£20.00 Hardback

256 pages, 80 black and white illustrations

978-1-4456-0057-4

Also abailable from Amberley Publishing

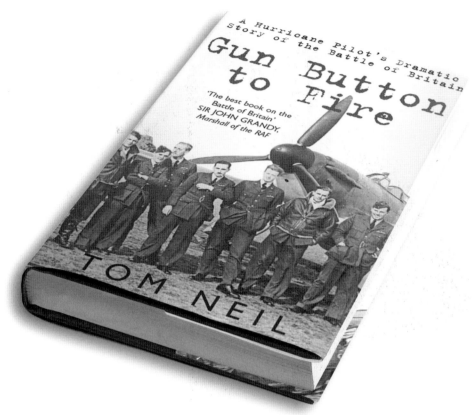

Gun Button To Fire
A Hurricane Pilot's Dramatic Story of the Battle of Britain

Tom Neil, DFC

This is a fighter pilot's story of eight memorable months from May to December 1940. When the Germans were blitzing their way across France, Pilot Officer Tom Neil had just received his first posting – to 249 Squadron. Nineteen years old, fresh from training at Montrose on Hawker Audax biplanes he was soon to be pitchforked into the maelstrom of air fighting on which the survival of Britain was to depend. By the end of the year he had shot down 13 enemy aircraft, seen many of his friends killed, injured or burned, and was himself a wary and accomplished fighter pilot.

£20.00 Hardback

329 pages, 120 illustrations (20 colour)
978-1-84868-848-3

Available from all good bookshops or to order direct
Please call **01285 760 030**
www.amberleybooks.com